Library of
Davidson College

VOID

COMPUTER ALGEBRA

PURE AND APPLIED MATHEMATICS

A Program of Monographs, Textbooks, and Lecture Notes

EXECUTIVE EDITORS

Earl J. Taft
Rutgers University
New Brunswick, New Jersey

Zuhair Nashed
University of Delaware
Newark, Delaware

CHAIRMEN OF THE EDITORIAL BOARD

S. Kobayashi
University of California, Berkeley
Berkeley, California

Edwin Hewitt
University of Washington
Seattle, Washington

EDITORIAL BOARD

M. S. Baouendi
Purdue University

Donald Passman
University of Wisconsin-Madison

Jack K. Hale
Brown University

Fred S. Roberts
Rutgers University

Marvin Marcus
University of California, Santa Barbara

Gian-Carlo Rota
Massachusetts Institute of Technology

W. S. Massey
Yale University

David Russell
University of Wisconsin-Madison

Leopoldo Nachbin
Centro Brasileiro de Pesquisas Físicas and University of Rochester

Jane Cronin Scanlon
Rutgers University

Anil Nerode
Cornell University

Walter Schempp
Universität Siegen

Mark Teply
University of Wisconsin-Milwaukee

LECTURE NOTES

IN PURE AND APPLIED MATHEMATICS

1. *N. Jacobson*, Exceptional Lie Algebras
2. *L.-Å. Lindahl and F. Poulsen*, Thin Sets in Harmonic Analysis
3. *I. Satake*, Classification Theory of Semi-Simple Algebraic Groups
4. *F. Hirzebruch, W. D. Newmann, and S. S. Koh*, Differentiable Manifolds and Quadratic Forms (out of print)
5. *I. Chavel*, Riemannian Symmetric Spaces of Rank One (out of print)
6. *R. B. Burckel*, Characterization of C(X) Among Its Subalgebras
7. *B. R. McDonald, A. R. Magid, and K. C. Smith*, Ring Theory: Proceedings of the Oklahoma Conference
8. *Y.-T. Siu*, Techniques of Extension on Analytic Objects
9. *S. R. Caradus, W. E. Pfaffenberger, and B. Yood*, Calkin Algebras and Algebras of Operators on Banach Spaces
10. *E. O. Roxin, P.-T. Liu, and R. L. Sternberg*, Differential Games and Control Theory
11. *M. Orzech and C. Small*, The Brauer Group of Commutative Rings
12. *S. Thomeier*, Topology and Its Applications
13. *J. M. Lopez and K. A. Ross*, Sidon Sets
14. *W. W. Comfort and S. Negrepontis*, Continuous Pseudometrics
15. *K. McKennon and J. M. Robertson*, Locally Convex Spaces
16. *M. Carmeli and S. Malin*, Representations of the Rotation and Lorentz Groups: An Introduction
17. *G. B. Seligman*, Rational Methods in Lie Algebras
18. *D. G. de Figueiredo*, Functional Analysis: Proceedings of the Brazilian Mathematical Society Symposium
19. *L. Cesari, R. Kannan, and J. D. Schuur*, Nonlinear Functional Analysis and Differential Equations: Proceedings of the Michigan State University Conference
20. *J. J. Schäffer*, Geometry of Spheres in Normed Spaces
21. *K. Yano and M. Kon*, Anti-Invariant Submanifolds
22. *W. V. Vasconcelos*, The Rings of Dimension Two
23. *R. E. Chandler*, Hausdorff Compactifications
24. *S. P. Franklin and B. V. S. Thomas*, Topology: Proceedings of the Memphis State University Conference
25. *S. K. Jain*, Ring Theory: Proceedings of the Ohio University Conference
26. *B. R. McDonald and R. A. Morris*, Ring Theory II: Proceedings of the Second Oklahoma Conference
27. *R. B. Mura and A. Rhemtulla*, Orderable Groups
28. *J. R. Graef*, Stability of Dynamical Systems: Theory and Applications
29. *H.-C. Wang*, Homogeneous Branch Algebras
30. *E. O. Roxin, P.-T. Liu, and R. L. Sternberg*, Differential Games and Control Theory II
31. *R. D. Porter*, Introduction to Fibre Bundles
32. *M. Altman*, Contractors and Contractor Directions Theory and Applications
33. *J. S. Golan*, Decomposition and Dimension in Module Categories
34. *G. Fairweather*, Finite Element Galerkin Methods for Differential Equations
35. *J. D. Sally*, Numbers of Generators of Ideals in Local Rings
36. *S. S. Miller*, Complex Analysis: Proceedings of the S.U.N.Y. Brockport Conference
37. *R. Gordon*, Representation Theory of Algebras: Proceedings of the Philadelphia Conference
38. *M. Goto and F. D. Grosshans*, Semisimple Lie Algebras
39. *A. I. Arruda, N. C. A. da Costa, and R. Chuaqui*, Mathematical Logic: Proceedings of the First Brazilian Conference

40. *F. Van Oystaeyen*, Ring Theory: Proceedings of the 1977 Antwerp Conference
41. *F. Van Oystaeyen and A. Verschoren*, Reflectors and Localization: Application to Sheaf Theory
42. *M. Satyanarayana*, Positively Ordered Semigroups
43. *D. L. Russell*, Mathematics of Finite-Dimensional Control Systems
44. *P.-T. Liu and E. Roxin*, Differential Games and Control Theory III: Proceedings of the Third Kingston Conference, Part A
45. *A. Geramita and J. Seberry*, Orthogonal Designs: Quadratic Forms and Hadamard Matrices
46. *J. Cigler, V. Losert, and P. Michor*, Banach Modules and Functors on Categories of Banach Spaces
47. *P.-T. Liu and J. G. Sutinen*, Control Theory in Mathematical Economics: Proceedings of the Third Kingston Conference, Part B
48. *C. Byrnes*, Partial Differential Equations and Geometry
49. *G. Klambauer*, Problems and Propositions in Analysis
50. *J. Knopfmacher*, Analytic Arithmetic of Algebraic Function Fields
51. *F. Van Oystaeyen*, Ring Theory: Proceedings of the 1978 Antwerp Conference
52. *B. Kedem*, Binary Time Series
53. *J. Barros-Neto and R. A. Artino*, Hypoelliptic Boundary-Value Problems
54. *R. L. Sternberg, A. J. Kalinowski, and J. S. Papadakis*, Nonlinear Partial Differential Equations in Engineering and Applied Science
55. *B. R. McDonald*, Ring Theory and Algebra III: Proceedings of the Third Oklahoma Conference
56. *J. S. Golan*, Structure Sheaves over a Noncommutative Ring
57. *T. V. Narayana, J. G. Williams, and R. M. Mathsen*, Combinatorics, Representation Theory and Statistical Methods in Groups: YOUNG DAY Proceedings
58. *T. A. Burton*, Modeling and Differential Equations in Biology
59. *K. H. Kim and F. W. Roush*, Introduction to Mathematical Consensus Theory
60. *J. Banas and K. Goebel*, Measures of Noncompactness in Banach Spaces
61. *O. A. Nielson*, Direct Integral Theory
62. *J. E. Smith, G. O. Kenny, and R. N. Ball*, Ordered Groups: Proceedings of the Boise State Conference
63. *J. Cronin*, Mathematics of Cell Electrophysiology
64. *J. W. Brewer*, Power Series Over Commutative Rings
65. *P. K. Kamthan and M. Gupta*, Sequence Spaces and Series
66. *T. G. McLaughlin*, Regressive Sets and the Theory of Isols
67. *T. L. Herdman, S. M. Rankin, III, and H. W. Stech*, Integral and Functional Differential Equations
68. *R. Draper*, Commutative Algebra: Analytic Methods
69. *W. G. McKay and J. Patera*, Tables of Dimensions, Indices, and Branching Rules for Representations of Simple Lie Algebras
70. *R. L. Devaney and Z. H. Nitecki*, Classical Mechanics and Dynamical Systems
71. *J. Van Geel*, Places and Valuations in Noncommutative Ring Theory
72. *C. Faith*, Injective Modules and Injective Quotient Rings
73. *A. Fiacco*, Mathematical Programming with Data Perturbations I
74. *P. Schultz, C. Praeger, and R. Sullivan*, Algebraic Structures and Applications Proceedings of the First Western Australian Conference on Algebra
75. *L. Bican, T. Kepka, and P. Nemec*, Rings, Modules, and Preradicals
76. *D. C. Kay and M. Breen*, Convexity and Related Combinatorial Geometry: Proceedings of the Second University of Oklahoma Conference
77. *P. Fletcher and W. F. Lindgren*, Quasi-Uniform Spaces
78. *C.-C. Yang*, Factorization Theory of Meromorphic Functions
79. *O. Taussky*, Ternary Quadratic Forms and Norms
80. *S. P. Singh and J. H. Burry*, Nonlinear Analysis and Applications
81. *K. B. Hannsgen, T. L. Herdman, H. W. Stech, and R. L. Wheeler*, Volterra and Functional Differential Equations

82. *N. L. Johnson, M. J. Kallaher, and C. T. Long*, Finite Geometries: Proceedings of a Conference in Honor of T. G. Ostrom
83. *G. I. Zapata*, Functional Analysis, Holomorphy, and Approximation Theory
84. *S. Greco and G. Valla*, Commutative Algebra: Proceedings of the Trento Conference
85. *A. V. Fiacco*, Mathematical Programming with Data Perturbations II
86. *J.-B. Hiriart-Urruty, W. Oettli, and J. Stoer*, Optimization: Theory and Algorithms
87. *A. Figa Talamanca and M. A. Picardello*, Harmonic Analysis on Free Groups
88. *M. Harada*, Factor Categories with Applications to Direct Decomposition of Modules
89. *V. I. Istrățescu*, Strict Convexity and Complex Strict Convexity: Theory and Applications
90. *V. Lakshmikantham*, Trends in Theory and Practice of Nonlinear Differential Equations
91. *H. L. Manocha and J. B. Srivastava*, Algebra and Its Applications
92. *D. V. Chudnovsky and G. V. Chudnovsky*, Classical and Quantum Models and Arithmetic Problems
93. *J. W. Longley*, Least Squares Computations Using Orthogonalization Methods
94. *L. P. de Alcantara*, Mathematical Logic and Formal Systems
95. *C. E. Aull*, Rings of Continuous Functions
96. *R. Chuaqui*, Analysis, Geometry, and Probability
97. *L. Fuchs and L. Salce*, Modules Over Valuation Domains
98. *P. Fischer and W. R. Smith*, Chaos, Fractals, and Dynamics
99. *W. B. Powell and C. Tsinakis*, Ordered Algebraic Structures
100. *G. M. Rassias and T. M. Rassias*, Differential Geometry, Calculus of Variations, and Their Applications
101. *R.-E. Hoffmann and K. H. Hofmann*, Continuous Lattices and Their Applications
102. *J. H. Lightbourne, III, and S. M. Rankin, III*, Physical Mathematics and Nonlinear Partial Differential Equations
103. *C. A. Baker and L. M. Batten*, Finite Geometries
104. *J. W. Brewer, J. W. Bunce, and F. S. Van Vleck*, Linear Systems Over Commutative Rings
105. *C. McCrory and T. Shifrin*, Geometry and Topology: Manifolds, Varieties, and Knots
106. *D. W. Kueker, E. G. K. Lopez-Escobar, and C. H. Smith*, Mathematical Logic and Theoretical Computer Science
107. *B.-L. Lin and S. Simons*, Nonlinear and Convex Analysis: Proceedings in Honor of Ky Fan
108. *S. J. Lee*, Operator Methods for Optimal Control Problems
109. *V. Lakshmikantham*, Nonlinear Analysis and Applications
110. *S. F. McCormick*, Multigrid Methods: Theory, Applications, and Supercomputing
111. *M. C. Tangora*, Computers in Algebra
112. *D. V. Chudnovsky and G. V. Chudnovsky*, Search Theory: Some Recent Developments
113. *D. V. Chudnovsky and R. D. Jenks*, Computer Algebra
114. *M. C. Tangora*, Computers in Geometry and Topology

Other Volumes in Preparation

COMPUTER ALGEBRA

Edited by

DAVID V. CHUDNOVSKY
Columbia University
New York, New York

RICHARD D. JENKS
IBM Thomas J. Watson Research Center
Yorktown Heights, New York

MARCEL DEKKER, INC.　　　　　　　　　　New York and Basel

Library of Congress Cataloging-in-Publication Data

Computer algebra / edited by David V. Chudnovsky. Richard D. Jenks.
    p.   cm. -- (Lecture notes in pure and applied mathematics :
  113)
    Papers of a conference held at the Courant Institute of
Mathematical Sciences of New York University. Apr. 5-6, 1984, and
sponsored by the National Science Foundation, and others.
    Includes bibliographies and index.
    ISBN 0-8247-8038-8
    1. Algebra--Data processing--Congresses.   I. Chudnovsky. D.
(David).   . II. Jenks, Richard D.   III. National Science
Foundation (U.S.) IV. Series: Lecture notes in pure and applied
mathematics ; v. 113.
QA155.7.E4C649 1989
512 .028 5--dc19

This book is printed on acid-free paper.

COPYRIGHT © 1989 by MARCEL DEKKER, INC.   ALL RIGHTS RESERVED

Neither this book nor any part may be reproduced or transmitted
in any form or by any means, electronic or mechanical, including
photocopying, microfilming, and recording, or by any information
storage and retrieval system, without permission in writing from
the publisher.

MARCEL DEKKER, INC.
270 Madison Avenue, New York, New York 10016

Current printing (last digit):
10 9 8 7 6 5 4 3 2 1

PRINTED IN THE UNITED STATES OF AMERICA

Preface

An international conference on Computer Algebra as a Tool for Research in Mathematics and Physics took place April 5-6, 1984, at the Courant Institute of Mathematical Sciences of New York University. This conference was organized to expose applications of computer algebra in pure and applied mathematics to a wide audience of scientists. The purpose was to expand the horizon of existing algebraic computational systems and their applications so that they might become useful, accessible, and effective tools of mathematical research. It was also designed to encourage discussions and suggestions that could lead to strengthening of the mathematical power of new and evolving computer algebra systems.

The organizing committee of the conference consisted of Gregory V. Chudnovsky, James Davenport, Jacob T. Schwartz, Stephen Wolfram, Shmuel Winograd, and David Y. Y. Yun. The co-chairmen were David V. Chudnovsky and Richard D. Jenks. The local arrangements chairman at New York University was Michael Overton. The conference was sponsored by the National Science Foundation, IBM, Inference Corporation, System Development Foundation, and Symbolics, Inc. Among the cosponsors were the American Physical Society, Association for Computing Machinery SIGSAM (Special Interest Group in Symbolic and Algebraic Manipulation), and SAME (Symbolic and Algebraic Manipulation in Europe).

Because of the large number of participants, the conference's lectures were conducted in the historic Great Hall of the Cooper Union in New York. Parallel with the conference, interactive demonstrations of existing computer algebra systems were conducted at New York University. Among the systems exhibited were MACSYMA (Symbolics, Inc.), MAPLE (University of

Waterloo, Canada), mu MATH (Soft Warehouse), REDUCE (RAND Corporation), SCRATCHPAD (IBM), SMP (Inference, Inc.), and CAYLEY (University of Sydney, Australia).

Invited lectures covered a large variety of areas of computer algebra systems, as well as development of the computer algebra systems themselves. The invited one-hour speakers included, in the order of their lectures, Anthony C. Hearn, James H. Davenport, William Gosper, Stephen Wolfram, Richard D. Jenks, George E. Andrews, Charles Sims, David B. Mumford, and William P. Thurston. Sessions were chaired by S. R. S. Varadhan, David Y. Y. Yun, Richard Fateman, Harvey Cohn, L. Ehrenpreis, and John W. Milnor. The closing session was devoted to the panel discussion, "The Potential of Computer Algebra as a Research Tool," the transcript of which is included in this volume. The panel consisted of Richard A. Askey, Michael E. Fisher, John McCarthy, and Joel Moses, and was moderated by Jacob T. Schwartz. The discussion was summarized by J. Barkley Rosser.

The number of registered participants exceeded 400. The conference attracted a wide spectrum of scientists. The organization skills of Jacob T. Schwartz, Michael Overton, and the conference secretary Tiyo Asai were crucial to the success of the conference. Most of the conference preparation rested with Richard D. Jenks.

We want to thank all sponsors of the conference, all invited speakers, and members of the organizing committee and the panel for their contributions.

This volume, long overdue, consists of contributions prepared especially for this conference, including the texts of invited talks and invited papers.

We hope that the conference has been a contributing factor in the growth of interest, use, and development of computer algebra and symbolic manipulation.

David V. Chudnovsky
Richard D. Jenks

Contents

PREFACE	iii
CONTRIBUTORS	ix

USE OF COMPUTER ALGEBRA FOR DIOPHANTINE AND
DIFFERENTIAL EQUATIONS 1
David V. Chudnovsky and Gregory V. Chudnovsky

1. Computer Algebra and Problems of Diophantine Approximations and Diophantine Geometry in the Function Field Case 2
2. Extremal Methods in the Theory of Diophantine Approximations 19
3. Diophantine Approximations to Various Numbers 34
4. Modular Equations and Approximations to π 40
5. Padé Approximations to Generalized Hypergeometric Functions of Type II 47
6. Measures of Diophantine Approximations for Values of Special G-Functions 55
7. The Arithmetic and Analytic Asymptotics of Padé Approximants 64
8. Differential Algebra Computations 75
   References 78

COMPUTER ALGEBRA AND THE HILBERT MODULAR EQUATION 83
Harvey Cohn

1. Introduction 83
2. The Weber Modular Equation 84
3. Hilbert Modular Functions and Equations 85
4. Theoretical Motivations 87
   References 88

SOME CAYLEY EXAMPLES 89
Michael C. Slattery

1. Introduction 89
2. Semidirect Products 90
3. A Small p-Group Example 94
4. Conclusion 95
   References 96

PHYSICS, RAMANUJAN, AND COMPUTER ALGEBRA 97
George E. Andrews

1. Introduction 97
2. The Rogers-Ramanujan Identities and the
   Hard Hexagon Model 98
3. The Rogers-Bailey Approach to the Rogers-
   Ramanujan Identities 102
4. Göllnitz's Theorem 105
5. Conclusion 108
   References 108

POLYPAK: AN ALGEBRAIC PROCESSOR FOR COMPUTATIONS
IN CELESTIAL MECHANICS 111
Dieter S. Schmidt

1. Introduction 111
2. Description of POLYPAK 113
3. Research Projects for Which POLYPAK Has Been Used 118
   References 119

COMPUTER ALGEBRA AND DEFINITE INTEGRALS 121
Richard Askey

ALGEBRAIC COMPUTATIONS AND STRUCTURES 129
James H. Davenport

1. Data Types of Computer Algebra 129
2. Constructive Algebra? 135
   References 143

COMPUTATION OF GALOIS GROUPS FROM POLYNOMIALS OVER
THE RATIONALS 145
David J. Ford and John McKay

1. Introduction 145
2. The Method 146
3. An Example 149
   References 150

ON THE USE OF SCRATCHPAD IN THE CONSTRUCTION OF
CONVOLUTION ALGORITHMS 151
Louis Auslander and Allan J. Silberger

1. Introduction 151
2. Computing Cyclic Convolution on Three Points 152
3. Cyclic Convolution on Nine Points 156
   Appendix 159
   References 181

AUTOMATED GENERATION OF OPTIMIZED CONVOLUTION ALGORITHMS *James W. Cooley*	183
1. Introduction	183
2. The Uses of SCRATCHPAD-Generated Convolution Algorithms	185
3. Mathematical Background	186
4. Derivation of the N = 7 Algorithm	188
5. Post-SCRATCHPAD Processing	194
6. A Suggested Extension of SCRATCHPAD	195
7. Conclusions	196
References	196
MANUAL FOR THE SYSTEM PNCRE *Robert Riley*	199
1. History of P	201
2. Preparation of the P System	205
3. Writing Programs that Call P Subroutines	209
4. The Sample Main Program KNOTS3	220
References	223
PANEL DISCUSSION	225
INDEX	239

Contributors

GEORGE E. ANDREWS Pennsylvania State University, University Park, Pennsylvania

RICHARD ASKEY University of Wisconsin-Madison, Madison, Wisconsin

LOUIS AUSLANDER Graduate Center of the City University of New York, New York, New York

DAVID V. CHUDNOVSKY Columbia University, New York, New York

GREGORY V. CHUDNOVSKY Columbia University, New York, New York

HARVEY COHN City College of the City University of New York, New York, New York

JAMES W. COOLEY IBM Thomas J. Watson Research Center, Yorktown Heights, New York

JAMES H. DAVENPORT University of Bath, Bath, England

DAVID J. FORD Concordia University, Montreal, Quebec, Canada

JOHN McKAY Concordia University, Montreal, Quebec, Canada

ROBERT RILEY State University of New York at Binghampton, Binghampton, New York

DIETER S. SCHMIDT University of Cincinnati, Cincinnati, Ohio

ALLEN J. SILBERGER Cleveland State University, Cleveland, Ohio, and IBM Thomas J. Watson Research Center, Yorktown Heights, New York

MICHAEL C. SLATTERY Marquette University, Milwaukee, Wisconsin

Use of Computer Algebra for Diophantine and Differential Equations

DAVID V. CHUDNOVSKY and GREGORY V. CHUDNOVSKY   Columbia University, New York, New York

This contribution grew out of our fascination with computer algebra systems, their potential, and their current power to guide and assist a researcher through complex computations. This chapter is an expanded version of our talk in which we tried to give a complete exposition of results that can easily be formulated (or expressed by a compact expression, e.g., Ramanujan's identities of Section 4), but are rather cumbersome to prove without the use of the computer. The chapter deals mainly with diophantine approximations, including diophantine equations and functional methods needed for their study. This includes, in particular, the methods of differential algebra, especially Padé approximation techniques, which are used extensively.

We greatly benefited from our work with the Computer Algebra Group at IBM, Yorktown Heights, especially from our interaction with R. Jenks, B. Trager, and B. Sutor. We also want to thank S. Winograd and D. Yun for their kind attention. Our tools included IBM SCRATCHPAD (old and the new one) and S.M.P. system, to which we were introduced by S. Wolfram. Our heartfelt thanks to him for his advice.

1. COMPUTER ALGEBRA AND PROBLEMS OF DIOPHANTINE APPROXIMATIONS AND DIOPHANTINE GEOMETRY IN THE FUNCTION FIELD CASE

In this section we collect results on diophantine equations and inequalities (diophantine approximations) over function fields (mainly over rational functions). Many of these results and methods that are used to prove them arose from and are used in the diophantine geometry (over $\mathbb{Z}$) and diophantine approximations of algebraic numbers and classical transcendence constants. We start with the functional field case, where the existing facilities of computer algebra can conveniently be applied. The specialization of functional results is described in detail in Sections 2, 3, 5, and 7. In those and other examples the search for a better measure of irrationality for numbers starts with functions whose values these numbers are: logarithmic, inverse trigonometric, elliptic, algebraic, and so on, and the search for rational approximations to these numbers is replaced by the search for rational (Padé) approximations to functions. Thus a large part of our applied work on computer algebra systems was devoted to a study of Padé approximations.

These Padé, or local rational approximations to systems of functions are important in our work on diophantine approximations to functions and their values, and also in the study of nonlinear differential equations with complete integrability properties.

Padé [41] defined [n/m]th Padé approximation to a function $f(x) = \sum_{n=0}^{\infty} c_n x^n$ as a rational function

$$\frac{P_n(x)}{Q_m(x)}$$

such that the expansion of (the remainder function)

$$f(x) - \frac{P_n(x)}{Q_m(x)}$$

in powers of x does not have exponents less than n + m. For given n and m, the [n/m]th Padé approximation to f(x) can be determined using determinants of the Hankel type in c_i of sizes (n + m) × (n + m). More efficient methods of computation of Padé approximation are based on linear recurrences relating contiguous Padé approximations that were described by Padé (and earlier by Jacobi and others). These recurrences turn into classical three-term linear recurrences relating orthogonal polynomials

Algebra for Diophantine Equations

when one considers the diagonal, [n/n]th, or near-diagonal, [n - 1/n]th, Padé approximations [42]. These recurrences are crucial for diophantine approximations, and they determine the continued fraction expansions of functions and numbers. For applications in mathematical physics, recurrences are interpreted as a spectral finite difference (lattice) problem, with x being a spectral parameter (see Refs. 5, 6, 10, 16, and 17). We also present a more general picture of Padé-Hermite approximations to systems of functions (see also Section 5 for a dual definition).

DEFINITION 1.1 Let $f_1(x), \ldots, f_m(x)$ be functions given by formal power series expansions at $x = 0$. For nonnegative integers $n_1, \ldots, n_m$ there exist polynomials $P_i(x|n_1,\ldots,n_m)$ of degrees at most n_i in x: $i = 1, \ldots, m$, such that (the remainder) function

$$\sum_{i=1}^{m} P_i(x|n_1,\ldots,n_m) f_i(x)$$

has a zero at $x = 0$ of an order at least $\sum_{i=1}^{m}(n_i + 1) - 1$. The polynomials $P_i(x|n_1,\ldots,n_m)$ are called Padé approximants to $\{f_1(x),\ldots,f_m(x)\}$ with weights $n_1, \ldots, n_m$.

Following Ref. 34, we put contiguous Padé approximants into a square matrix

$$\phi(x|N_1,\ldots N_m) = (P_i(x|N_1 + \delta_{j1}, \ldots, N_m + \delta_{jm}))_{i,j=1}^{m}$$

Then the contiguous recurrence relations between Padé approximants take the form

$$\phi(x|N_1 + 1, \ldots, N_m + 1) = (xI + P(N_1,\ldots,N_m))\phi(x|N_1,\ldots,N_m) \quad (1.1)$$

for a scalar $m \times m$ matrix $P(N_1,\ldots,N_m)$ (and the "local transfer matrix" $xI + P(N_1,\ldots,N_m)$) (see Ref. 18).

The identity (1.1) shows how to compute Padé approximants recursively, when "next" Padé approximants are calculated from the preceding ones by solving, on every new step, only a few linear equations on the order of the remainder function. This method lends itself to a simple parallel version as well.

The methods of computer algebra were used, however, to obtain an easier method of determining local transfer matrices. This method is applicable to systems of functions $\{f_1(x),\ldots,f_m(x)\}$ satisfying linear differential equations with rational function coefficients. In this case, in addition to (1.1), we have differential equations

$$\frac{d}{dx} \phi(x|N_1,\ldots,N_m) = A_N(x)\phi(x|N_1,\ldots,N_m) \tag{1.2}$$

The combination of (1.1) and (1.2) allows us to compute consecutively the local transfer matrices relating contiguous Padé approximants, starting from rational function coefficients entering differential equations for $f_1(x), \ldots, f_m(x)$. The corresponding (nonlinear) recurrence formulas are known as Bäcklund transformations, and are entirely different from a traditional way of computing Padé approximants starting from power series expansions of functions. They are also quite efficient (see Ref. 43).

One of the crucial problems of Padé approximations and of convergence of Padé approximations is the problem of normality. Padé approximations to functions $f_1(x), \ldots, f_m(x)$ are called *normal* if for arbitrary weights $n_1, \ldots, n_m$ the remainder function

$$\sum_{i=1}^{m} P_i(x|n_1,\ldots,n_m) f_i(x)$$

has a zero at $x = 0$ of order *exactly*

$$\sum_{i=1}^{m} (n_i + 1) - 1$$

A more realistic condition is that of *almost normality*, when there is a constant $C \geq 0$ (depending only on $f_1, \ldots, f_m$) such that the order of zero of

$$\sum_{i=1}^{m} P_i(x|n_1,\ldots,n_m) f_i(x)$$

is at most $\sum_{i=1}^{m}(n_i + 1) - 1 + C$.

This problem is related to an old problem of Kolchin [44] on rational approximations to solutions of (algebraic) differential equations. The Kolchin problem asks whether for arbitrary $P(x)$, $Q(x)$, and a solution of $f(x)$ of a differential equation,

$$\text{ord}_{x=0}\left[f(x) - \frac{P(x)}{Q(x)}\right] \leq \deg(P) + \deg(Q) + \varepsilon(\deg(P) + \deg(Q))$$

whenever $\varepsilon > 0$ and $\deg(P) + \deg(Q) \geq c_1(\varepsilon, f)$. Although this is much weaker than the almost-normality property, Kolchin's problem is of immediate

Algebra for Diophantine Equations

importance because of its clear relation to the problem of Rost-type estimates in number theory.

Kolchin was also the first to point out the relation of this problem to differential algebra and differential invariants. It is here, in the solution of the Kolchin problem, that computer algebra was crucial in getting the initial insight and in considerable simplification of proofs. First, here is the result (an extremely simple, two-page proof of which is given in Ref. 45):

THEOREM 1.2 [45,46] Let $f_1(x), \ldots, f_m(x)$ be arbitrary solutions of linear differential equations [over $\mathbb{C}(x)$] regular at $x = 0$. Then there exists a constant $C_2 > 0$ (depending only on differential equations satisfied by $f_1, \ldots, f_m$) such that for arbitrary polynomials $P_1(x), \ldots, P_m(x)$ we have

$$\mathrm{ord}_{x=0}[P_1(x)f_1(x) + \cdots + P_m(x)f_m(x)]$$
$$\leq \sum_{i=1}^{m} \deg(P_i) + C_2 \left[\sum_{i=1}^{m} \deg(P_i)\right]^{1/2} + C_2$$

In particular, for every $\varepsilon > 0$,

$$\mathrm{ord}_{x=0}\left(\sum_{i=1}^{m} P_i f_i\right) \leq (1 + \varepsilon) \sum_{i=1}^{m} \deg(P_i)$$

if $\sum_{i=1}^{m} \deg(P_i) \geq C_2'(\varepsilon)$.

For the proof of this theorem we had to construct a sequence of differential polynomials $D_N(P_1, \ldots, P_m)$ in undetermined $P_1, \ldots, P_m$ with coefficients from $\mathbb{C}[x]$, of the degree in each $P_1, \ldots, P_m$ at most α_N, such that

$$\mathrm{ord}_{x=0}[D_N(P_1, \ldots, P_n)] \geq \beta_N \, \mathrm{ord}_{x=0}\left[\sum_{i=1}^{m} P_i(x)f_i(x)\right] - C_3(N)$$

and such that $\alpha_N/\beta_N \to 1$ as $N \to \infty$.

We conducted a computer search for such differential invariants using SCRATCHPADs and SMP for $m = 2$ and the second-order linear differential equations. This helped us to identify (almost immediately) differential polynomials D_N having the best $1 - \alpha_N/\beta_N$ as a particular wronskian formula that appeared earlier in our work in nonlinear completely integrable systems (Ref. 47), where they represented multisoliton-like solutions. Namely, $D_N(P_1, \ldots, P_m)$ can be taken as a ratio of wronskians:

$$D_N(P_1,\ldots,P_m) = \frac{W(P_i\varphi_\alpha : \alpha \in A_N, \; i = 1,\ldots,m)}{W(\varphi_\alpha : \alpha \in A_N)^m}$$

where $\varphi_\alpha : \alpha \in A_N$ is the basis of solutions of a (scalar) ordinary linear differential equation satisfied by all functions $f_1^{n_1}\ldots f_m^{n_m} : n_1 + \cdots + n_m = N$. With this form of D_N it is obvious how to complete the proof of Theorem 1.2.

Theorem 1.2 is far from the best possible. In fact, we conjecture that *every* system of functions satisfying linear differential equations with rational function coefficients is *almost normal*. We have already some progress in the direction of this conjecture, the most significant of which is the following [10]:

THEOREM 1.3 Padé approximations to arbitrary algebraic functions are almost normal.

The same is true for large classes of solutions of fuchsian linear differential equations. In particular, in the continued fraction expansion of an algebraic function, its partial fractions have bounded degrees. This (nonarchimedean) functional result is in striking contrast with the number-theoretic situation, when for a nonquadratic algebraic number we expect an unbounded continued fraction expansion. We have to leave to future computer algebra systems an interesting and important problem of bounding the constant C in the almost-normality property in terms of the number of branch points (singularities) of $f_i(x)$.

One of the exciting areas of future applications of computer algebra is diophantine geometry, particularly the solution of diophantine equations. In this field numerical methods have played and are still playing an important role, but can be enhanced with the use of computer algebra systems.

As an example of our own research, we look at the classical problem of effective determination of integer points on algebraic curves. Here Siegel's theorem [1] implies that there are only finitely many integer points whenever a curve is nonrational. It is natural to consider in this respect a crucial example of the so-called Thue equations

$f(X,Y) = A$

for a binary form $f(X,Y) = Y^n f(X/Y,1)$ of degree $n \geq 3$, with integer coefficients. Here A is an integer, and X and Y are integral solutions. There are, essentially, three known methods of effective solution of Thue equations:

(i) Skolem's method based on p-adic expandions of solutions of exponential equations; (ii) Baker's method based on lower bounds of linear forms in logarithms of algebraic numbers; and (iii) Padé approximation methods related to Thue-Siegel results (in this respect, see particularly Section 3).

Recently, we significantly enlarged the class of diophantine equations (and inequalities) that can be handled using Padé-type approximation techniques [10-12,14-15,40]. Our techniques are based on Padé-type approximations to G-functions (see Section 6 for definitions and some of the results), including algebraic functions, using such methods as graded Padé approximations [10,12]. These techniques are closely related to diophantine geometry over function fields, and in the application of computer algebra, we deal now with effective solutions of functional diophantine equations. Here we use our wronskian methods in conjunction with Picard-Fuchs linear differential equations and isomonodromy deformation equations associated with them (see Ref. 10).

The first important class of functional diophantine equations is given by Norm-form equations for function fields. Let K be an arbitrary function field which is a finite extension of $\mathbb{C}(x)$ of degree $n \geq 3$. Let $m < n$ and let $f_1 = f_1(x),\ldots,f_m = f_m(x)$ be linearly independent over $\mathbb{C}$ elements of K, which do not generate a nontrivial subfield of K.

The Norm-form equations (or, correctly, "noncomplete Norm-form equations") have the form

$$\text{Norm}_{K/\mathbb{C}(x)}(X_1 f_1 + \cdots + X_m f_m) = A \tag{1.3}$$

for a fixed polynomial $A = A(x) \in \mathbb{C}[x]$ and polynomial unknowns $X_i = X_i(x)$: $i = 1, \ldots, m$. The *ineffective* Schmidt's theorem [13] (in both functional and number-theoretical cases) asserts that there are only finitely many polynomial solutions of (1.3). Note that for $m = n = [K : \mathbb{C}(x)]$, the Norm-form equations have typically infinitely many solutions.

In the functional case we obtain an effective bound on (degrees of) solutions $X_i = X_i(x)$ of (1.3) which is close to the best possible:

THEOREM 1.4   Let $\varepsilon > 0$, $n > m$, and let $A = A(x)$ be a polynomial of degree D. Then the degrees of polynomial solutions $X_i = X_i(x)$: $i = 1, \ldots, m$ of (1.3) are bounded by

$$\frac{1 + \varepsilon}{n - m} D + \gamma_0(K,\varepsilon)$$

where $\gamma_0(K,\varepsilon)$ is an effective constant whose dependence on K is linear in terms of degrees over $\mathbb{C}(x)$ of coefficients of algebraic equations, defining $f_1, \ldots, f_m$ and K.

For earlier results, see Refs. 47 and 48. For $n \geq 3$, $m = 2$ (the Thue equation case) the constant $\gamma_0(K,\varepsilon)$ can be computed using existing computer algebra systems.

The statement of Theorem 1.4, if translated for the number-theoretical case, would mean that solutions of Norm-form equations, in particular, Thue equations, are always bounded *polynomially* in heights of coefficients of the equation. This is a fascinating conjecture, going contrary to the existing bounds supplied, say, by Baker's method. The functional result makes this conjecture plausible. In fact, our extensive numerical experiments support this conjecture. Moreover, examples of cubic and quartic irrationalities show that there is analytic evidence in favor of this conjecture (modulo various L-function conjectures) along with the experimental evidence.

Another interesting class of functional diophantine equations is given by the Mordell conjecture. In the functional case, the solution of Mordell's conjecture is effective [49] and Siegel's theorem on integer points can easily be made effective. We use for these purposes Manin's method [49] relying on Picard-Fuch's equations satisfied by periods of algebraic curves as functions of parameters in the equations of curves. It is interesting to compare this functional technique with a similar p-adic technique related to the p-curvature operator that had been studied in our work on the Grothendieck conjecture for rank 1 equations over algebraic curves of genus $g \geq 1$ [38].

We present one result giving an upper bound on integer points on elliptic curves defined over function fields. In the proof of this result, computer algebra methods were used to determine certain nonlinear differential expressions from nonconstant points on elliptic curves and, especially, precise orders of vanishing of wronskians at fibers of bad reductions according to Kodaira classification. The nonlinear differential expression in question is Manin's map μ [49]. This map μ is a homomorphism of the set $E_{\mathbb{C}(t)}$ of all $\mathbb{C}(t)$-rational points on (an elliptic) curve E over $\mathbb{C}(t)$ into $\mathbb{C}(t)$:

$$\mu: E_{\mathbb{C}(t)} \mapsto \mathbb{C}(t)$$

Algebra for Diophantine Equations

For example, if an elliptic curve E is taken in the Legendre form $E_\lambda : y^2 = x(x-1)(x-\lambda)$, then for an arbitrary point $P(\lambda) = (X(\lambda), Y(\lambda))$ on E_λ, the expression for $\mu(P(\lambda))$ can be determined from the following Painlevé equation (viii), from the Painlevé list of equations without movable critical points [50]:

$$\frac{d^2 X}{d\lambda^2} = \frac{1}{2}\left(\frac{1}{X} + \frac{1}{X-1} + \frac{1}{X-\lambda}\right)\left(\frac{dX}{d\lambda}\right)^2 - \left(\frac{1}{\lambda} + \frac{1}{\lambda-1} + \frac{1}{X-\lambda}\right)\frac{dX}{d\lambda} + \frac{X(X-1)}{2\lambda(\lambda-1)(X-\lambda)} + \mu(P(\lambda))Y$$

Analysis of various bad fibers gives us [10]

THEOREM 1.5   Let $E : y^2 = 4x^3 + g_2(t)x + g_3(t)$ be an elliptic curve over $\mathbb{C}[t]$ with polynomials $g_2(t)$, $g_3(t)$ such that $\Delta(t) = g_2(t)^3 - 27 g_3(t)^2$ is not identically zero. Then for an integer (polynomial) point $(X(t), Y(t))$ on E, the degree of $X(t)$ is bounded by

$$\deg(X(t)) \leq \max\left\{\deg(g_2), \frac{\deg(g_3)}{2}, 6\deg(\Delta(t)) + c\right\}$$

for an absolute constant $c > 0$.

The most important applications of new G-function theorems (see Section 5 and Refs. 10, 12, and 14) are to algebraic functions that are obviously G-functions (via the Eisenstein theorem). This way we can treat a very interesting problem of effective solution of Norm-form equations depending on parameters. The class of equations we consider are Norm-form equations of the form

$$\mathrm{Norm}_{K/\mathbb{Q}}(X_1 \alpha_1 + \cdots + X_m \alpha_m) = A \tag{1.4}$$

where $\alpha_1, \ldots, \alpha_m$ are linearly independent over $\mathbb{Q}$ elements of the algebraic number field K, and $A \in \mathbb{Z}$.

Schmidt's theorem [13] states that for a fixed A, equation (1.4) can have infinitely many integer solutions $X_1, \ldots, X_m$ only when in a $\mathbb{Z}$-submodule of K generated by $\alpha_1, \ldots, \alpha_m$ there exist a submodule M, and a subfield L of K, such that $M = \lambda L$ for $\lambda \in K^*$ and for a $\mathbb{Z}$-submodule L of L of rank $= [L : \mathbb{Q}]$, where L is not a quadratic imaginary field (nor $\mathbb{Q}$).

Only for $m = 2$ (Thue equations) are there Baker's effective bounds on solutions of (1.4). However, these bounds are always exponential in sizes of coefficients of the left side of (1.4) and far from the best possible in $|A|$. Schmidt [13] presents (ineffective) upper bounds on solutions of (1.4) in terms of $|A|$ for a primitive field K (i.e., having no nontrivial subfields). The effectivizations of Schmidt's or even Thue-Siegel's theorems are far from being achieved. Nevertheless, in the functional case, results close to the best possible were obtained in Theorem 1.4.

Inspired by successful functional results, we look at their number-theoretical analog when a parameter is present: it is an integer parameter on which coefficients of the Norm-form equation depend:

$$P(X_1,\ldots,X_m;N) \ [= \text{Norm}(X_1\alpha_1 + \cdots + X_m\alpha_m)] = A \tag{1.5}$$

That is, α_i: $i = 1, \ldots, m$ are functions of N. In the case when all α_i are nonsingular at $N = \infty$ (with some restrictions on the field of definition of coefficients of their expansion at $N = \infty$), we can apply G-function theorems [see Refs. 10, 12, and 14 and Section 6] to bound integral solutions of (1.5) for large $|N|$. These bounds are close to the best possible in $|A|$ for large $|N|$. They are always polynomial in $|N|$ and $|A|$. Moreover, with some notable (rationally parametrized families of) exceptions, solutions of (1.4) can be bounded uniformly in $X_1, \ldots, X_m$ and N; that is, for a *fixed* A, there are only *finitely many* integral solutions $(X_1,\ldots,X_m,N)$ of (1.5).

To reduce a solution of equation (1.4) to the problem of diophantine approximations of values of algebraic functions, we represent the equations (1.4) and (1.5) in the form

$$\text{Norm}_{K/\mathbb{Q}(x)}[X_1\alpha_1(x) + \cdots + X_m\alpha_m(x)] = A \tag{1.6}$$

for a function field $K \supseteq \mathbb{Q}(x)$ and algebraic functions $\alpha_1(x), \ldots, \alpha_m(x) \in K$ and $x = N$. The equation (1.6) can be decomposed into

$$\prod_{i=1}^{d} [X_1\alpha_1^{(i)}(x) + \cdots + X_m\alpha_m^{(i)}(x)] = A \tag{1.7}$$

where $d = [K : \mathbb{Q}(x)]$, and, in applications, $x = N$ is a (rational) integer.

In the formulation of (1.7) we are ready to apply G-function theorems. For example, if all branches $\alpha_1^{(i)}(x), \ldots, \alpha_m^{(i)}(x)$ are nonsingular at $x = \infty$, with rational coefficients of Taylor expansions, then Theorem 6.6 (see Ref. 14) can be applied directly and yields in this situation a uniform effective bound of solutions of equations (1.5)-(1.7) for fixed A, which is very close to the best possible (in fact, upper bounds on solutions are close to lower bounds existing for certain parametrized families of solutions).

Unfortunately, in many cases branches $\alpha_j^{(i)}(x)$ can be singular at $x = \infty$, even when $\alpha_j(x)$ are regular at $x = \infty$.

In the notations of (1.7), we are reducing the problem of solution of (1.5) to a problem of simultaneous diophantine approximations of numbers $\alpha_1^{(i)}(N), \ldots, \alpha_m^{(i)}(N)$, since for a fixed A, we have

$$\prod_{i=1}^{d} |X_1 \alpha_1^{(i)}(N) + \cdots + X_m \alpha_m^{(i)}(N)| \sim O(1)$$

and for $m < d$, at least one of the forms

$$|X_1 \alpha_1^{(i)}(N) + \cdots + X_m \alpha_m^{(i)}(N)|$$

is very small with respect to

$$X = \max\{|X_1|, \ldots, |X_m|\}$$

for large X. See Ref. 13 for rigorous arguments in a more general and complex situation.

Hence we arrive at the problem of a simultaneous diophantine approximation to values of algebraic functions. In the case $m = 2$ it is relatively easy to get an answer. One of our main results on the finiteness of a number of solutions of a Thue equation depending on a parameter is the following:

THEOREM 1.6 Let us consider a <u>Thue equation</u> depending on an integer parameter N:

$$f(X,Y;N) \stackrel{\text{def}}{=} Y^n F\left(N, \frac{X}{Y}\right) = A \tag{1.8}$$

where $F(x,y) \in \mathbb{Z}[x,y]$ is irreducible over $\mathbb{Q}[x,y]$. Let all real branches $y = y(x)$ of $F(x,y) \equiv 0$ have power series expansions at $x = \infty$ with integral exponents and rational coefficients (or coefficients all lie in a field of degree $< n$). Then there are at most finitely many rationally parametrized solutions of (1.8). These parametrized solutions have the form

$$\frac{Y}{X} = \frac{P(N)}{Q(N)} \qquad \text{A fixed}$$

for $P(x), Q(x) \in \mathbb{Q}[x]$. These parametrized solutions can be determined as exceptionally good rational approximations $P(x)/Q(x)$ of a real branch $y = y(x)$ of $F(x,y) \equiv 0$ at $x = \infty$, that violate the near-normality of Padé approximations—there are only *finitely many* such exceptionally good approximations.

With the exception of parametrized solutions, Thue equation (1.8) *has only finitely many integral solutions* (X,Y;N) for a fixed A. Moreover, the bound of solutions of (1.8) is close to the best possible in $|A|$. For any $\varepsilon > 0$, and $|N| \geq N_1(\varepsilon)$, the nonparametrized solutions X, Y of (1.8) are bounded from above as follows:

$$\max\{|X|,|Y|\} \leq \gamma_1(\varepsilon)|A|^{1/(n-2)-\varepsilon}$$

Here $\gamma_1(\varepsilon) > 0$ is an effective constant depending on $\varepsilon > 0$ and *polynomially* on the height of a polynomial $F(x,y)$.

All exceptional rationally parametrized (in N) solutions of (1.8) can be found by looking at the continued fraction expansion of various real branches $y = y(x)$ of $F(x,y) \equiv 0$, and looking through the initial part of the expansion. There exists an effective upper bound for the exceptionally good approximation (their degrees are bounded, in view of our theorem on almost normality of Padé approximations to algebraic functions [10]).

The statement of Theorem 1.6 is incorrect when there are branches $y = y(x)$ of $F(x,y) \equiv 0$ that are real at $x = \infty$ and have *nontrivial* Puiseux expansion. In this case one can observe the appearance of nontrivial unirationally parametrized families of solutions of (1.8): when X, Y, and N are polynomials in a new integral variable t.

Similar results can be proved for general Norm-form equations:

THEOREM 1.7 Let us consider a Norm-form equation depending on an integral parameter N:

$$P(X_1,\ldots,X_m;N) \stackrel{\text{def}}{=} \text{Norm}_{K/\mathbb{Q}(N)}[X_1\alpha_1(N) + \cdots + X_m\alpha_m(N)]$$

$$= \prod_{i=1}^{d}[X_1\alpha_1^{(i)}(N) + \cdots + X_m\alpha_m^{(i)}(N)] = A \quad (1.9)$$

for algebraic functions $\alpha_1(x), \ldots, \alpha_m(x)$ in a function field of degree $d \geq 3$ over $\mathbb{Q}(x)$. Let us assume that *all* branches $\alpha_i^{(j)}(x)$ ($i = 1, \ldots, m$; $j = 1, \ldots, d$) of algebraic functions have power series expansions at $x = \infty$ with integral exponents and rational coefficients (or all coefficients lie in an algebraic field of degree < d). Let us assume that K does not have nontrivial subfields [or, weaker, that the module generated by $\alpha_1(x), \ldots, \alpha_m(x)$ is not equivalent to the full module in a nontrivial subfield of K], and that $\alpha_1(x), \ldots, \alpha_m(x)$ are linearly independent over $\mathbb{C}(x)$. Then there are only finitely many parametrized solutions of (1.9) with a given A:

Algebra for Diophantine Equations

$$X_1 = X_1(N), \ldots, X_m = X_m(N)$$

for $X_1(x), \ldots, X_m(x) \in \mathbb{Q}[x]$. With the exception of these rationally parametrized solutions, equation (1.9) *has only finitely many integral solutions* $(X_1, \ldots, X_m; N)$ for a fixed A. Moreover, if $\varepsilon > 0$ and $|N| \geq N_2(\varepsilon)$, the nonparametrized solutions $X_1, \ldots, X_m$ of (1.9) are bounded from above as follows:

$$\max\{|X_1|, \ldots, |X_m|\} \leq \gamma_2(\varepsilon) |A|^{1/(d-m) - \varepsilon}$$

This bound is the best possible (up to $-\varepsilon$). Moreover, the last bound is effective and so are $N_2(\varepsilon)$ and $\gamma_2(\varepsilon)$. The result of the theorem is not effective, however, because for $|N| < N_2(\varepsilon)$ one has to invoke Schmidt's theorem [13] to obtain the finiteness result.

One of the other examples of the finiteness of a number of solutions of a Thue equation depending on a parameter refers to the famous Runge theorem [51] as its starting point:

THEOREM 1.8   Let $F(x,y) \in \mathbb{Z}[x,y]$ be a polynomial irreducible over $\mathbb{Q}[x,y]$, $F(x,y) = f_n(x,y) + f'_{<n}(x,y)$, where $f_n(x,y)$ is homogeneous of degree $n \geq 3$, $f'_{<n}$ consists of monomials of degree $< n$, and $f_n(1,y)$ is reducible over $\mathbb{Q}$ but does not have multiple roots. The associated Thue equations, depending on an integer parameter N,

$$f(N;X,Y) \stackrel{\text{def}}{=} Y^n F\left(N, \frac{X}{Y}\right) = A$$

have only *finitely many* (effectively determined) solutions in triplets (N,X,Y) of integers, for any given A, with the possible exception of finitely many families of rationally parametrized solutions. They are given by

$$\frac{X}{Y} = \frac{P_0(N)}{Q_0(N)}$$

for rational functions $P_0(x)/Q_0(x)$ whose degrees and coefficients are uniformly bounded.

Here $P_0(x)/Q_0(x)$ can be effectively determined as Padé approximations to an algebraic function $y = y(x)$ in $F(x,y) \equiv 0$ violating normality condition. (See Theorem 1.3 on near-normality of Padé approximations to algebraic functions.)

In Theorem 1.8 the trivial (hyperbolic) case of $A = 0$ corresponds to Runge's theorem [51]. For nonparametrized solutions, the upper bounds in terms of $|A|$ are the best, of Roth's type. For example, if $\varepsilon > 0$ and $|N| \geq N_0(\varepsilon)$, solutions X and Y of a Thue equation (1.8) are bounded as $\max\{|X|,|Y|\} \leq \gamma_1(\varepsilon)|A|^{1/(n-2)+\varepsilon}$.

EXAMPLES 1.9 (a) The diophantine equations

$$(N + 1)X^n - NY^n = A$$

have only finitely many solutions

$$(N,X,Y,n)$$

in integers (and $n \geq 3$), whenever A is fixed.

(b) The equations

$$X^3 - NXY^2 + Y^3 = 1 \qquad N \leq 1$$

have families of "parametrized" solutions, (N,X,Y):

$$(N,1,N),(N,1,0),(N,0,1) \tag{1.10}$$

With these exceptions there are only finitely many integer solutions (N,X,Y). They are (N,X,Y)

$$(1,-1,1),(1,4,3),(-1,-2,3)$$

In these examples, the check of all *possible* parametrized solutions (exceptionally good rational approximations to algebraic functions) was run on SCRATCHPAD, and small nonparametrizable solutions were checked numerically.

(c) The equation

$$X^{2k+1} - NXY^{2k} + Y^{2k+1} = 1 \qquad N < 0$$

does not have integral solutions (N,X,Y,k) for $k \geq 2$, with the exception of rationally parametrized solutions (1.10).

One of the targets of our applications was the study of Thus equations depending on a parameter of the form

$$x^n - Nxy^{n-1} + y^n = A$$

$N,A \in \mathbb{Z}$ in integers x and y. The corresponding algebraic functions α, $\alpha^n - N\alpha + 1 = 0$, have expansions at $N = \infty$:

$$\alpha = \alpha_0(N) \sim N^{-1} + \cdots$$

$$\alpha = \alpha_j(N) \sim e^{2\pi i j/(n-1)} N^{1/(n-1)} + \cdots$$

$j = 0, \ldots, n - 2$.

Our main result on the "$2 + \varepsilon$" exponent of the measure of irrationality for G-functions (Refs. 10, 14, and 15) immediately implies that the exponent of irrationality of $\alpha = \alpha_0(N)$ is "$2 + \varepsilon$" whenever $|N| \geq N_3(\varepsilon,n)$.

In particular, we have a very good uniform bound for solutions of the Thue equation above whenever $N < 0$ and only $\alpha_0(N)$ is a real root for odd n.

Some time ago we looked at the case of $n = 3$ when explicit Padé approximations to $\alpha_0(N)$ can be constructed. One of the applications of our method is the following bound for solutions x and y in terms of N and A (see Ref. 10, Chap. 11, and Ref. 40): If $x^3 - Nxy^2 + y^3 = A$ and x, y are distinct from $\pm 1, 0$, then for $N < 0$ and any $\varepsilon > 0$,

$$|A| \geq \gamma_4(\varepsilon)|N||x|^{2+\chi-1}$$

and

$$\chi = \frac{\pi + \sqrt{3} \log(16|N|^3/27)}{\pi - \sqrt{3} \log(16|N|^3/27)}$$

and $\chi \to -1$ as $|N| \to \infty$. Here $\gamma_4(\varepsilon)$ is an effective constant > 0.

The case of $N > 0$ is more complicated, and in this case there are additional rational and unirational parametrized families of solutions as well as more exceptional cases. We present results in the case of a positive discriminant as follows: For $N > 0$ there are the following classes of solutions of a Thue equation

$$x^3 - Nxy^2 + y^3 = 1:$$

$(x,y;N)$: $(1,0;N),(0,1;N),(1,N;N),(x,1;x^2),(1 - y^3, y; y^4 - 2y)$ and $(x,y;N)$: $(-1,1;1),(4,-3;1)$; $(-2,-3;2),(-1,-2;2)$; $(-1,-1;3),(-3,-2;3),(2,-1;3)$; $(508,273;4)$; $(-2,-1;5),(7,-3;5)$; $(18,7;7),(-19,7;7)$; $(-26,9;8)$; $(28,9;10)$; $(-7,2;12)$; $(13,3;19)$; $(-14,-3;22)$; $(-791,114;48)$.

The problem of a solution in triplets $(X,Y;N)$ of a (cubic) Thue equation

$$f_3(X,Y;N) \stackrel{\text{def}}{=} Y^3 F_3\left(N, \frac{X}{Y}\right) = A$$

is, in fact, equivalent to the problem of finding integral points on an elliptic *surface*

$$f(X,Y;N) = A \qquad (1.11)$$

The problem of determining all those N for which (1.11) has (nontrivial) integral solutions is clearly a part of this task. It is remarkable that certain conditions on $F_3(x,y)$ can imply the uniform boundedness (finiteness) of solutions (X,Y;N) of (1.11). No similar result is proved, say, for other elliptic curves with complex multiplications, depending on a parameter N:

$$x^3 - y^2 = N.$$

In any case, our results, like the boundedness of solutions (X,Y,N) of

$$X^3 - NXY^2 + Y^3 = 1$$

(with the exception of rationally and unirationally parametrized families of solutions), are first results for the finiteness of the number of integral points on *surfaces* and in this case go beyond both functional and number-theoretic versions of Siegel's theorem. They also strongly suggest that the bounds on integral points on algebraic curves (of positive genus) should be polynomial in heights of the defining equation.

The problem of the number of solutions of diophantine equations (as opposed to the problem of actually finding the solutions) also leads to the use of Padé-type approximation methods, since it turns out that knowledge of one or several integral solutions introduces certain integral parameters in the problem, around which one can expand algebraic functions. Cubic Thue equations have been under investigation since 1919 [52]. For cubic-form equations

$$f(x,y) \stackrel{\text{def}}{=} ax^3 + bx^2y + cxy^2 + dy^3 = 1$$

and $a,b,c,d \in \mathbb{Z}$, two methods were derived for determination of the number of solutions. One, due to Delone-Paddeev-Nagel [52], treats the case of a negative discriminant $D < 0$. In this case there are at most five solutions: five solutions when $D = -23$, four when $D = -44$ and $D = -31$, and at most three in all other cases. Three solutions exist for the rationally parametrized case: $x^3 - nxy^2 + y^3 = 1$: $(x,y;n) \sim (1,0;n), (0,1;n), (1,n;n)$.

For $D > 0$ Siegel proved in 1929 [1] using Padé approximations to $z^{1/3}$ that there are at most 18 solutions for sufficiently large discriminants. In fact, Siegel's method gives an upper bound of 10 for the number of solutions with sufficiently large D (Gelman, see Ref. 52). Similarly, Siegel's method gives an upper bound of 12 for all discriminants $D > 0$ (J-H. Evertse). In fact, we have

THEOREM 1.10   For sufficiently large $D \geq D_0$, the cubic form equation $f(x,y) = 1$ has at most nine solutions.

Probably, this result is true for all $D > 0$ and is the best possible. We do know at least one cubic form with nine solutions. This is the form with $D = 49$, solved by Baulin: $x^3 + x^2 y - 2xy^2 - y^3 = 1$ with solutions $(x,y) \sim (1,0),(0,-1),(-1,-1),(-1,1),(2,-1),(-1,2),(5,4),(4,-9),(-9,5)$. It seems that there should be infinitely many cubic forms corresponding to cyclic fields for which the cubic-form equation has nine solutions.

All the results above correspond to method (iii) noted above for the study of diophantine approximations to algebraic numbers using Padé-type approximation techniques. The two other methods of effective solution of diophantine equations cited above also benefit from the use of computer algebra systems.

The Skolem method (i) is based on the p-adic expansion of the zeros of such exponential equations as

$$\alpha_1 \varepsilon_1^x + \cdots + \alpha_n \varepsilon_n^x = 0$$

for algebraic numbers α_i, ε_i: $i = 1, \ldots, n$. Limited applications of Skolem's method can be expanded by means of power series programs and symbolic manipulations with them.

Baker's method [(ii); Ref. 3] is another field for the application of computers and computer algebra. One of the interesting problems here is the explicit determination of all constants in lower bounds of linear forms in logarithms of algebraic numbers. This is a typical problem for computer algebra, for Baker's method of proof consists of solution of systems of algebraic equations and inequalities depending on parameters. We ran a test program of this type on SCRATCHPAD for linear forms in two logarithms of rational numbers and came up with the following lower bond:

$$|k^{j/n} - \ell| > \exp\left\{-e^{10} \frac{\ln H(k) \ln H(\ell)}{\ln(1 + k_*)} \left[1 + \frac{\ln n}{\ln(1 + k_*)}\right]^2\right\}$$

where k and ℓ are (multiplicatively independent) rational numbers, $k > 1$, $|1 - k| \leq k_*^{-1}$ for $k_* > 1$, and n and j are positive integers, $n \geq j$ (see Ref. 2, Chap. 5).

Improvements in Baker's method are important because of their various applications, say, in the problem of effective determination of all elliptic curves with given points of bad reduction. The latter problem is quite useful for the case-by-case verification of the Weil conjecture. According to this conjecture, for every N, an elliptic curve, defined over $\mathbb{Q}$ and having a conductor N, is uniformized by automorphic functions corresponding to a congruence subgroup $\Gamma_0(N)$ of the full modular group $SL_2(\mathbb{Z})$. The Weil conjecture itself is only a starting point in a series of conjectures on elliptic curves, their L-functions, and integer and rational points on the curves. All the conjectures benefit from the new and expanded large-scale computer algebra systems. The Weil conjecture and its various generalizations give an alternative (fourth) approach to the effectivization of the Thue-Siegel-Roth theorem. We will sketch an approach to this problem that uses modular forms in the case of cubic irrationalities. First, as explained above, diophantine approximations to cubic irrationalities can be substituted by the diophantine equation $f_3(x,y) = A$ for a cubic form f_3. Second, one can apply Mordell's reduction (see Ref. 40) and derive an equation for invariants G_2 and G_3 of the elliptic curve: $4G_2^3 - 27G_3^2 = D(27A)^2$, where $G_2 = 3H(x,y)$, $G_3 = G(x,y)$, H and G are quadratic and cubic covariants of $f_3(x,y)$, and D is the discriminant of f_3. We look now at an elliptic curve $E_{x,y} : Y^2 = X^3 - G_2 X - G_3$ defined over $\mathbb{Z}$, depending on an integral solution (x,y) of $f_3(x,y) = A$. Now $\Delta = D(27A)^2$ is its discriminant (in this Weierstrass form)—depending only on f_3 and A, and to the modular invariant $J = 12^3 \cdot 4G_2^3/\Delta$ there corresponds τ in the fundamental domain of $SL_2(\mathbb{Z})$ such that $J = J(\tau)$. If we assume Weil's conjecture, then $E = E_{x,y}$ is parametrized by modular functions on $\Gamma_0(N)$, where N is a conductor of E (e.g., N consists of primes dividing Δ raised to bounded powers according to the Ogg-Shafarevich formula). The (normalized) differential of the first kind on E is given by a modular form $f(\tau)$ of weight 2 on $\Gamma_0(N)$. To bound G_2 and G_3 (i.e., to bound $|x|$ and $|y|$ in terms of D and A), one has to bound ω and ω' of E. In fact, for $|J| > 1$, it is enough to bound only one of the periods. To get this bound, we suggest looking at Patterson's inner product $<f,f> = \iint_{H/\Gamma_0(N)} |f(\tau_R + i\tau_I)|^2 \, d\tau_R \, d\tau_I$ of $f(\tau)$. It is well known that $<f,f>$ is a rational multiplier of $\omega\omega'$ (the area of the fundamental cell of the period lattice of E). This rational factor can be estimated for a given N using either group-theoretic methods (looking at representations of $SL_2(\mathbb{Z})/\Gamma_0(N)$), or looking at the action of Frobenius on the jacobian of $\Gamma_0(N)$. Simple bounds obtained this way give an upper bound on $|x|$, $|y|$ in $f_3(x,y) = A$ in terms of exponentials in N. It is natural to assume that one can prove (unconditionally) better estimates of $<f,f>/\omega\omega'$ in terms of quantities polynomial in N. This would imply the best possible bounds on the solutions of the Thue equation $f_3(x,y) = A$ in both the archimedean and p-adic cases in terms of A and the discriminant of f_3. More generally, the Weil conjecture provides us with a means of proving Hall's conjecture [53], according to which $|x^2 - y^3| \geq c \max\{|x|^{1/3}, |y|^{1/2}\}$.

2. EXTREMAL METHODS IN THE THEORY OF DIOPHANTINE APPROXIMATIONS

Analytic methods in the theory of diophantine approximations to the values of a function with "good" arithmetic properties are based on the construction of systems of approximations to functions themselves. Results on the arithmetic nature of values of these functions (their irrationality, transcendence, and measures of rational approximations) are deduced by specialization of these approximating systems. Although the arithmetic properties of functions for which nontrivial results are proved vary significantly (e.g., from entire functions with Siegel's E-function property [1] to elliptic (abelian) logarithms [2]), the common pattern is easily discerned. One requires good convergence both in the archimedean (complex) and nonarchimedean domains in the neighborhood of a sequence of points related with the point of approximation. (In the case of E-functions there is only one point of approximation where "good" convergence properties are guaranteed. This is $x = 0$. For elliptic and abelian logarithms, one looks at approximations determined by local matrices in the neighborhood of a point on an abelian variety. The same is true in the degenerate case, for example, for linear forms in logarithms of algebraic numbers [3].) Padé approximations are a natural tool to be used in the construction of the desired systems of approximations; specifically, one talks about multipoint Hermite-Padé approximations [4], determined (often uniquely) by the maximal allowable number of conditions on the degrees and orders of approximation at given points. This approach is extremely efficient in cases when explicit systems of Padé approximations are known [5,6]. An example of this is the case of a binomial (or logarithmic) function, where one-point Padé approximants were constructed by Hermite. In this particular case archimedean and nonarchimedean convergence of Padé approximants was determined completely in Refs. 6-7. Another example is that of Hermite-Padé approximations to generalized hypergeometric functions that we describe in detail here, following the initial exposition in Ref. 9. Unfortunately, in most of the interesting cases, closed expressions of Padé approximations are not known, and moreover, requirements of archimedean and nonarchimedean convergence, necessary for applications, are not satisfied generically even for the simplest classes of functions. This concerns essentially any solution of a linear differential equation with more than three (regular) singularities, not reducible to the hypergeometric case. An extreme example is a simple quadratic algebraic function $y(x) = \sqrt{P_4(x)}$ for $P_4(x) \in \mathbb{Z}(x)$, where Padé approximants to $y(x)$ at $x = \infty$ have denominators of coefficients growing as c^{n^2}, where n is the weight of Padé approximations whenever $x = \infty$ is not a torsion point of an elliptic curve $y^2 = P_4(x)$. This, together with the need to improve measures of irrationality of numbers that are values of functions whose Padé approximations are known, leads to a need to improve upon approximations given by Padé approximations. One way to do it is to investigate more complicated schemes of Padé approximations involving multivariable generalizations and the construction of "graded Padé approximations" [10-12]. Also, to prove many general results one does not need Padé approximations but rather, Padé-type approximations, where the number of approximation conditions (in terms of degrees and orders of Padé approximants) is significantly less than the maximal allowable. In this approach one uses "ineffective" constructions, where Padé-type approximations are not constructed but are proved to exist with explicit bounds on the sizes of their coefficients. (This "ineffective" method of proof is not to be confused with ineffective results of the Thue-Siegel-Roth theorem [3,13], where the approximation system is proved to exist—provided that several initial approximations to algebraic numbers are known—to show that these approximations do not actually exist. In the case of "ineffective" proofs, all the results are still effective.) Padé-type approximations were particularly effective in the proof of general results for values of the G-function; see Refs. 10, 12, and 14. For example, for algebraic functions, particularly good convergent Padé-type approximations were constructed (see Ref. 15) in the notorious case of $\sqrt{P_4(x)}$ described above. Unfortunately, for particular numbers, such as $\ln 2$, π, $\sqrt[3]{5}$, ..., most of the "ineffective" proofs do not give good results

in view of the inflated bounds on archimedean and p-adic convergence, as
guaranteed by existence results. This makes it necessary to look for effective realizations of Padé-type approximations in the classes of functions that include solutions of fuchsian linear differential equations.
One approach to an effective construction of new Padé-type approximations
was studied by the authors from the point of view of Bäcklund transformations and contiguous relations arising from the isomonodromy deformation
of linear differential equations [5,6,16-18]. In this approach ordinary
Padé approximations were determined from the local multiplicities at regular singularities, while Padé-type approximations were determined from
isomonodromy deformations with apparent singularities. Several explicit
examples of such an approach, together with recurrences connecting consecutive approximants, were presented in Refs. 17-19 for particular values of
logarithmic and inverse trigonometric functions such as ln 2, $\pi/\sqrt{3}$, and π.
See the references cited above for explicit expressions of Padé-type approximants. Simultaneously with this general approach we realized that
one can often express solutions of appropriate linear differential equations in a closed form as integrals of rational or algebraic functions.
In fact, many of the linear differential equations that are satisfied by the
remainder function and Padé-type approximants are Pochhammer equations
solvable in quadratures. This led to an investigation of an optimal form
of integrand in Pochhammer-type integral representation that provides us
with the best Padé-type approximations for values of logarithmic, binomial,
and inverse trigonometric functions. An optimal choice of the integrand
is equivalent to the extremal problem of the best rational (polynomial)
approximation on a given interval (continuum) of a rational function.
This is a traditional problem in approximation theory, where one has to
impose new arithmetic extremality conditions: integrality (or near-integrality with a fixed set of possible bad primes and controllable orders
at these primes) of polynomials involved. This problem of mixed analytic-arithmetic extremality of polynomials was analyzed in detail in Ref. 19
in connection with the closely related problem of elementary methods in
the prime number theorem. Formulas from Ref. 19 based on integral equations are important in our study of diophantine approximations. Another
important component is the WKB method, which often reduces the estimate
of the integrals $\int_\gamma Q(\bar{x})^N d\bar{x}$ as $N \to \infty$ to a study of critical points of
$Q(\bar{x})$. We present below a variety of applications of the study of integral
representation of this kind. A key element in our study was the ability

to determine with high precision critical points of integrands of Pochhammer type. This was achieved by means of traditional (IMSL) and entirely new high-precision packages for polynomial root finding developed by us.

The necessary analytic tool in the construction of a variety of approximations to arithmetically important functions is the method of integral representation of periods of algebraic varieties. We provide the necessary details in the one-dimensional case, where our main object is the study of "periods" of abelian integrals of the form

$$\int_\gamma \prod_{i=1}^n (\zeta - a_i)^{\alpha_i - 1} \, d\zeta$$

where γ is a fixed path (on the Riemann surface of the integrand). [In the case of rational exponents α_i, this integral indeed represents periods of (possibly degenerate) abelian integrals.]

The class of such integrals are known as Pochhammer integrals. In his study Pochhammer [20] chose one of a_i as a variable, say, z, and investigated the dependence of the integral on z. Let us put forth his notations:

$$W(\alpha_1, \ldots, \alpha_m, \mu; z) = \int_C (\zeta - z)^{\mu + n - 1} \prod_{i=1}^m (\zeta - a_i)^{\alpha_i - 1} \, d\zeta \qquad (2.1)$$

where the contour C is one of two kinds: either (i) C is a closed path not containing any of a_i or z; or (ii) C is a path beginning and ending at points where the integrand vanishes. The function in (2.1) is the Euler representation of a solution of a linear ordinary differential equation with polynomial coefficients. In fact, if we define two polynomials Q(z) and P(z) as

$$Q(z) = \prod_{i=1}^m (z - a_i)$$

$$\frac{P(z)}{Q(z)} = \sum_{i=1}^m \frac{\alpha_i}{z - a_i} \qquad (2.2)$$

then $W(z) = W(\alpha_i, \ldots, \alpha_m, \mu; z)$ [for all contours C as in (i) and (ii) above] satisfies the equation

$$\sum_{i=0}^m (-1)^{m-i} \binom{\mu}{i} Q^{(m-i)}(z) \frac{d^i}{dz^i} W = \sum_{i=0}^{m-1} (-1)^{m-1-i} \binom{\mu + 1}{i} P^{(m-1-i)}(z) \frac{d^i}{dz^i} W$$

(2.3)

Algebra for Diophantine Equations   23

There are natural bases of solutions of (2.3) corresponding to single or double loops C in (i). For i = 1, ..., m, we denote by γ_i a simple loop beginning and ending at a fixed base point (hereafter denoted by x_0) and circling a_i in the positive direction. By $-\gamma_i$ we denote the same loop described in the reverse direction. Similarly, we introduce the loop γ_z, encircling $\zeta = z$ in the ζ-plane. Let us denote by W_i a function (2.1), where $C = \gamma_i$, and for i,j = 1, ..., m we denote by W_{ij} a function in (2.1), where the contour C is $C = \gamma_i + \gamma_j - \gamma_i - \gamma_j$. Similarly, by W_{iz} we denote a function in (2.1) with $C = \gamma_i + \gamma_z - \gamma_i - \gamma_z$. The relationship between the loops gives the following basic representation:

$$W_{k\ell} = (1 - e^{2\pi i \alpha_\ell})W_k - (1 - e^{2\pi i \alpha_k})W_\ell \tag{2.4}$$

from which we deduce

$$(1 - e^{2\pi i \alpha_j})W_{k\ell} = (1 - e^{2\pi i \alpha_k})W_{\ell j} + (1 - e^{2\pi i \alpha_\ell})W_{kj}$$

$$(1 - e^{2\pi i \mu})W_{k\ell} = (1 - e^{2\pi i \alpha_k})W_{\ell z} + (1 - e^{2\pi i \alpha_\ell})W_{kz}$$

The monodromy group of equation (2.3) can be obtained formally from relations (2.4). To determine the generators of the monodromy group of (2.3), let us consider the effect on the basis of (2.3) of the analytic continuation along the closed path (from x_0 to x_0) encircling a single singularity a_j. The following linear transformations of fundamental solutions occur:

$$W_z \xmapsto{\gamma_j} W_z + e^{2\pi i \mu}W_{jz} \quad W_j \xmapsto{\gamma_j} -W_{jz} + W_j \quad W_k \xmapsto{\gamma_k} W_k \quad \text{for } k \neq j \tag{2.5}$$

and as a consequence of (2.4) and (2.5),

$$W_{jz} \to e^{2\pi i(\mu+\alpha_j)}W_{jz}; \quad W_{kz} \mapsto W_{kz} - e^{2\pi i\mu}(1 - e^{2\pi i \alpha_j})W_{jz} \quad (\text{for } k \neq j)$$

The Pochhammer equation together with other classes of equations whose solutions can be determined explicitly are the only cases where the monodromy group is known explicitly. Knowledge of the monodromy group allows us to construct, following the method of Refs. 5 and 6, Padé approximations and Padé-type approximations to solutions of Pochhammer-type equations (2.3). All Padé approximations again have the form (2.1) with different exponents α_i (differing from the original ones by integers) and with, possibly, additional apparent singularities a_i. These Padé and Padé-type approximants are related by contiguity formulas [which are linear relations between functions (2.1) when the exponents α_i change in integer increments]. In the case of functions (2.1), the contiguity relations can all be obtained from the following few basic relations.

For an arbitrary function of the form (2.1) with C of the form (i) or (ii), we get after differentiation of the integrand,

$$\frac{dW(\alpha_1,\ldots,\alpha_m,\mu;z)}{dz} = -(\mu + n - 1)W(\alpha_1, \ldots, \alpha_m, \mu - 1; z) \qquad (2.6\text{i})$$

Similar to (2.6i), one gets other identities replacing d/dz by d/da_i, $\mu + n$ by α_i. In addition to (2.6i), two more classes of basic contiguous relations exist:

$$W(\alpha_1 + 1, \alpha_2, \ldots, \alpha_m, \mu; z) = W(\alpha_1, \alpha_2, \ldots, \alpha_m, \mu + 1; z)$$
$$+ W(\alpha_1,\ldots,\alpha_m,\mu;z)(z - a_1) \qquad (2.6\text{ii})$$

$$(\mu + n - 1) \frac{\partial W}{\partial a_1} - (\alpha_1 - 1) \frac{\partial W}{\partial z} = (z - a_1) \frac{\partial^2 W}{\partial a_1 \partial z}$$

This allows us to express the functions $W(\alpha_1 + n_1, \ldots, \alpha_m + n_m, \mu + n; z)$ for integers n_i in terms of any m linearly independent functions of the same form (contiguous functions) with coefficients that are rational in a_i, α_i, z_i, and μ. The integral representation (2.1), differential equation (2.3), and contiguous relations (2.6) are major ingredients in the application of Pochhammer integrals to explicit construction of Padé-type approximations.

Further topics on Pochhammer integrals include the study of Picard's group of monodromy of these integrals in (m + 1)-dimensional complex space of parameters $a_1, \ldots, a_m$, z and its relation with Siegel's modular forms. This topic, including improvements on classical Picard's results, was the subject of recent important work of Deligne-Mostow [21].

Let us consider applications of Pochhammer-type integrals to diophantine approximations of values of the logarithmic and inverse trigonometric functions. We represent these integrals in the most general form,

$$I = \int_C \prod_\xi (\zeta - \alpha_\xi)^{\nu_\xi} d\zeta \qquad (2.7)$$

where α_ξ are algebraic numbers such that $\nu_\xi = \nu_{\xi'}$, whenever α_ξ and $\alpha_{\xi'}$ are algebraically conjugate, and C is a path of type (i) or (ii). Varying paths C, we obtain all analytic continuations of integrals in (2.7). Looking at the monodromy relations (2.4) and (2.5), one deduces that integrals in (2.7) are linear combinations of logarithmic functions iff all exponents v_ξ in (2.7) are *integers* (positive or negative). In fact, one can see that under these conditions and the foregoing assumptions on v_ξ, any integral I of the form (2.7), with $C = \overline{\alpha_{\xi_1} \alpha_{\xi_2}}$ being a path connecting two α_{ξ_1} and α_{ξ_2} for which v_{ξ_1} and v_{ξ_2} are positive, is a linear combination of logarithms, namely,

$$I = \left[\int_{\alpha_{\xi_1}}^{\alpha_{\xi_2}} \Pi \, (\zeta - \alpha_\xi)^{v_\xi} \, d\zeta \right] = \sum_{v_\xi < 0} \log \frac{\alpha_{\xi_2} - \alpha_\xi}{\alpha_{\xi_1} - \alpha_\xi} D_\xi + D_0 \qquad (2.8)$$

where D_ξ are algebraic numbers and such that $D_{\xi'}$ and D_ξ are algebraically conjugate whenever $\alpha_{\xi'}$ and α_ξ are algebraically conjugate. This simple observation is sufficient to deduce a variety of Padé approximations to logarithmic functions. Hermite's construction of Padé approximations to exponential, binomial, and logarithmic functions follows exactly the same outline (see Ref. 22). It is enough to look at two classes of examples.

EXAMPLE 2.1   In the Padé approximation problem to the system of functions $\log(1 - \omega_i t)$: $i = 1, \ldots, m$ (with distinct ω_i) at $t = 0$ one considers integrals of the form

$$I = \int_C \zeta^n (\zeta - 1)^n \prod_{i=1}^m \left(\zeta - \frac{1}{\omega_i t} \right)^{-n-1} d\zeta \qquad (2.9)$$

If one takes $C = \overline{01}$ in (2.9), I is the remainder function in the Padé approximation problem to $\log(1 - \omega_i t)$: $i = 1, \ldots, m$ at $t = 0$ with the weight n. The Padé approximants in this problem also have the form (2.9) for $C = \gamma_i$ - a loop around $\alpha_i = 1/\omega_i t$ for $i = 1, \ldots, m$.

The Padé approximation problem adjoint to Example 2.1 consists of simultaneous approximation to 1, $\log(1 - \omega_i t)$: $i = 1, \ldots, m$ at $t = 0$, that is, of systems of polynomials $Q_i(t)$: $i = 0, \ldots, m$ such that the remainder functions $R_i(t) \overset{\text{def}}{=} Q_i(t) - Q_0(t) \log(1 - \omega_i t)$ at $t = 0$ have zeros of orders $n + [n/m] + 1$: $i = 0, \ldots, m$, where degrees of polynomials $Q_i(t)$: $i = 0, \ldots, m$ are bounded by n. (These are the so-called Padé approximations of the second kind, and n is the weight of Padé approximations.) Similar to Example 2.1, these Padé approximations arise as follows:

EXAMPLE 2.2 We look at integrals

$$I = \int_C \prod_{i=1}^{m} \left(\zeta - 1 - \frac{1}{\omega_i t}\right)^n \frac{(\zeta - 1)^n}{\zeta^{n+1}} d\zeta \tag{2.10}$$

If $C = \overline{1 \cdot \alpha_i}$ for $\alpha_i = 1/\omega_i t$: $i = 1, \ldots, m$ we obtain the remainder functions $R_i(t)$ above (after multiplication by appropriate powers of t). The Padé approximants with weight n correspond to different choices of C.

Examples 2.1 and 2.2 were used by Hermite, and later they and their multidimensional generalizations (see below) were used by Mahler, Baker, and others (see references in Ref. 18) to obtain measures of irrationality of logarithms of rational (and algebraic) numbers. Particularly successful were the ordinary Padé approximations to ln(1 - t) corresponding to m = 1 in (2.9) or (2.10) with $\omega_1 = 1$, with Padé approximants and the remainder function expressed in terms of Legendre polynomials and Legendre functions of the first and second kind. Later, in Ref. 17-19, we developed Padé-type approximations to logarithmic functions giving better measures of diophantine approximations. These methods of new Padé-type approximations can now be deduced from appropriate Pochhammer integrals.

New methods of better approximations to ln h (**for** rational or algebraic h) are based on the study of integrals

$$I = \int_1^h \frac{\prod_\xi (\zeta - \alpha_\xi)^{v_\xi}}{\zeta^{n+1}} d\zeta \tag{2.11}$$

where $\alpha_0 = 1$, $\alpha_1 = h$; v_ξ are (nonnegative) integers and, as above, $v_\xi = v_{\xi'}$ whenever α_ξ and $\alpha_{\xi'}$ are algebraically conjugate. The integrals (2.11) thus can be represented in the form

$$I = \int_C \frac{(\zeta - 1)^{v_0}(\zeta - h)^{v_1} \prod_{k \geq 2} P_k(\zeta)^{v_k}}{\zeta^{n+1}} d\zeta \tag{2.11'}$$

where $P_k(\zeta) \in \mathbb{Z}[\zeta]$. The case of $v_k = 0$ for $k \geq 2$ corresponds to the usual Padé approximations $I = P_n(h) \ln h - Q_n(h)$ to ln h at h = 1 for I in (2.11). The extremality conditions that determine the integrals (2.11) giving the best rational approximations to ln h are determined by (a) how close I in (2.11) is to zero; (b) how far from 0 Padé-type approximants in (2.11') are; and (c) how small the common denominator of the coefficients of the polynomial Padé-type approximants is.

Algebra for Diophantine Equations

The answer to questions (a)-(c) for a *given* integrand in (2.11) and (2.11') can be obtained effectively from the WKB method [in the complex domain for parts (a) and (b) and in the p-adic domain for part (c)]. To make the formulation simple, let us assume (without any loss of generality) that all exponents v_ξ depend linearly on a single parameter n [in the denominator of (2.11) and (2.11')]. (In fact, one can prove that interesting results can be obtained *only* in this situation.) That is, we can assume that

$$v_\xi = [w_\xi n]: \quad \xi = 0, 1, \ldots$$

for fixed (rational) w_ξ, and that n is a sufficiently large integer.

As follows from the monodromy discussion with Example 2.2, the integral I has the form

$$I = P \ln h + Q \qquad (2.12)$$

where P and Q are rational functions in h (polynomials in α_ξ) with rational number coefficients. The "Padé-type approximants" can also be represented as (linear combinations of) integrals in (2.11') for various paths C. The expression of P in (2.12) is, in fact, very simple:

$$P = \mathrm{res}_{\zeta=1} \left[\frac{(\zeta - 1)^{v_0} (\zeta - h)^{v_1} \Pi_{k \geq 2} P_k(\zeta)^{v_k}}{\zeta^{n+1}} \right] \qquad (2.13)$$

Thus P is a polynomial in h and in coefficients of $P_k(\zeta)$ for $k \geq 2$.

We are able now to give a complete answer to the three-part question (a)-(c) of the analytic and arithmetic asymptotics of the remainder function I and "Padé-type approximants" P and Q in (2.12) as $n \to \infty$ for $v_\xi = [w_\xi n]$, as above.

To formulate results on the arithmetic asymptotics of (2.12) we assume that h is a rational (or even any algebraic) number. Then we have:

PROPOSITION 2.3 Let h and α_ξ be all algebraic numbers belonging to the field K. Let S be the set of all nonarchimedean valuations v of K such that $v(h) \neq 0$ or $v(\alpha_\xi) \neq 0$ for some ξ. If $M = \max\{n, \Sigma_\xi v_\xi - n\}$, and $\Delta_n = \mathrm{lcm}\{1, \ldots, M\}$, then numbers $\Delta_n P$ and $\Delta_n Q$ for P and Q from (2.12) are S-integral numbers from K. Moreover, for any $v \in S$, $\min\{v(P), v(Q)\} = \min\{\mu_1, \mu_2\}$, where $\mu_1 [= \mu_1(v)] = \Sigma_{\xi, v(\alpha_\xi) < 0} v_\xi v(\alpha_\xi)$, $\mu_2 [= \mu_2(v)] = \Sigma_{\xi, v(\alpha_\xi) < v(h)} [v(\alpha_\xi) - v(h)] v_\xi + v(h)(\Sigma_\xi v_\xi - n)$.

In particular, if P are prime ideals of K corresponding to v, then

$$D = \Delta_n \prod_{P, v_p(\alpha_\xi h) \neq 0} P^{-\min\{\mu_1(v_P), \mu_2(v_P)\}}$$

is the common denominator of $P, Q \in K$ from (2.12). The archimedean part of the height of P and Q, together with the asymptotics of $|I|$ in (2.12), is determined from the steepest descent method applied to (2.11) and (2.11'). The form of integrands and the Laplace method allows us to use the classical form of the steepest descent method, applied originally by Riemann [23] to integrals (2.11) and (2.11') with $v_k = 0$ for $k \geq 2$. For the modern exposition of this method and all the necessary auxiliary results (contour deformation, Watson lemma), see Ref. 24. A simple application of the steepest descent method leads to the following:

PROPOSITION 2.4 As above, let $v_\xi = [w_\xi n]$, $\xi = 0, 1, \ldots$ for fixed w_ξ and sufficiently large n, where α_ξ in (2.11) and (2.11') are independent of n. We put $R(\zeta) = \Pi_\xi (\zeta - \alpha_\xi)^{w_\xi} \zeta^{-1}$, and denote by z_j all critical points of $R(\zeta)$ [i.e., roots of the equation $\Sigma_\xi w_\xi / (\zeta - \alpha_\xi) = 1/\zeta$]. Then the asymptotics of integrals (2.11') are determined by the contribution of critical points z_j lying on the steepest descent paths homological to C. In particular, whenever I in (2.11') is not identically zero as $n \to \infty$,

$$\frac{1}{n} \log|I| \leq \max_{z_j} \log|R(z_j)| \tag{2.14}$$

where max is taken over all critical points z_j of $R(\zeta)$. On the other hand, if h is real > 1 and $w_0 > 0$, $w_1 > 0$, then

$$\frac{1}{n} \log|I| \leq \max_{z_j \in (1,h)} \log|R(z_j)| \tag{2.15}$$

where max is taken only over those critical points z_j that lie in the interval $(1, h)$.

The asymptotics of (2.15) hold, in fact, for arbitrary complex h not lying on the cut from $-\infty$ to 1. In this case one chooses as a path between 1 and h the steepest descent/ascent path.

REMARK One can explicitly determine conditions under which equalities in (2.14) and (2.15) are achieved. Moreover, if all critical points z_j are determined, all steepest descent/ascent paths determining I, P, and Q in (2.12) can be determined explicitly from solutions of a simple (one-particle) dynamical system governed by the integrand $R(\zeta)^n$ in the complex ζ-plane.

Algebra for Diophantine Equations

The key problem becomes a problem of determination of all critical points of $R(\zeta)$, that is, the solution of a polynomial equation of degree $\Sigma_{v_\xi > 0} 1$ with high accuracy. This is achieved by our specialized high-precision version of the parallel/vector polynomial root-finding package. The ability to determine for a given integrand $R(\zeta)^n$ all critical points of $R(\zeta)$ and the asymptotics of integrals in (2.14) and (2.15) allows us to solve the extremality problem and to find the best choice of exponents v_ξ and polynomials $P_k(\zeta) \in \mathbb{Z}[\zeta]$ in (2.11'). The figure of merit that determines the extremality condition is expressed in terms of the ratio of arithmetic and analytic asymptotics of $|I|$ and heights of P and Q in (2.12). The corresponding measure of irrationality is described by a simple lemma on dense approximations.

LEMMA 2.5   Let us assume that there exist a sequence of rational integers P_n, Q_n such that

$$\left.\begin{array}{l}\log|P_n|\\ \log|Q_n|\end{array}\right\} \sim an \quad \text{as } n \to \infty$$

and $\log|P_n \theta - Q_n| \sim bn$ as $n \to \infty$, where $b < 0$. Then the number θ is irrational and for any $\varepsilon > 0$ and for all rational integers p, q, we have

$$|\theta - \frac{p}{q}| > |q|^{a/b - 1 - \varepsilon} \qquad (2.16)$$

provided that $|q| \geq q_0(\varepsilon)$.

Thus one tries to find α_ξ and v_ξ such that for the sequence of approximations, $DI_n = (DP_n) \ln h + (DQ_n)$ from (2.12) satisfies the conditions of Lemma 2.5 with the minimal $|a/b|$.

Here a and b are determined from the asymptotics (2.14) and (2.15) as $n \to \infty$ from the contribution of critical points z_j of $R(\zeta)$; and from the asymptotics $1/n \log|D|$ of the common denominator $D \in \mathbb{Z}$ [for $h \in \mathbb{Q}$, $P_k(\zeta)$ $\mathbb{Z}[\zeta]$] as described explicitly in Proposition 2.3.

One sees that the extremality problem formulated this way does not have a unique solution whenever the number of distinct zeros α_ξ of the integrand is unbounded. When one bounds the number of α_ξ, the best form of the integrand can be (numerically) determined. One sees immediately that by increasing the number of zeros α_ξ, any bound on the exponent $a/b - 1$ in Lemma 2.5 for $\theta = \ln h$ can be improved. To improve the exponent it is enough to introduce a new factor $P_{k+1}(\zeta)^{v_{k+1}}$ into the integrand, whose roots are exactly the critical points of the integrand $R(\zeta)^n$. [By definition, the polynomial $P_{k+1}(\zeta) = R'(\zeta) \Pi_\xi (\zeta - \alpha_\xi)$ has rational number coefficients.] Consequently, any bound on the exponent of the measure of irrationality of ln h that can be obtained using integrals (2.11) and (2.11') can always be further improved, the only change being in the constant $q_0(\varepsilon)$ in the bound (2.16). (Hopefully, the exponent will tend to -2, although there is but a single example of h for which it can be proved.)

This process of iterative improvement can best be illustrated starting from the case of $v_k = 0$ for $k \geq 2$ corresponding to the usual Padé approximations $I_n = P_n(h) \ln h - Q_n(h)$ to ln h. The equation of the critical points of the integrand in this case is $\zeta^2 - h = 0$, giving rise to asymptotics of $(1/n) \log |I_n|$, $(1/n) \log |P_n(h)|$, or $(1/n) \log |Q_n(h)|$ given by $\log|1 - \sqrt{h}|^2$ and $\log|1 + \sqrt{h}|^2$, respectively. Thus the next obvious choice for integrals in (2.11) and (2.11') is

$$I = \int_1^h \frac{[(\zeta - 1)(\zeta - h)]^m (\zeta^2 - h)^{n-m}}{\zeta^{n+1}} d\zeta \qquad (2.17)$$

The new Padé-type approximants to logarithmic functions presented in Section 4 of Ref. 18 can, in fact, be identified using integral representation (2.17). In this case the integral I in (2.17) is a linear combination of 1 and ln h:

$$I = P_n(h) \ln h - Q_n(h) \qquad (2.18)$$

where

$$P_n(h) = \sum_{i=0}^{m} \sum_{\substack{j=0 \\ n-m \leq i+2j \leq n}}^{n-m} \binom{m}{i} \binom{m}{n-i-2j} \binom{n-m}{j} (-1)^j h^{i+j} \qquad (2.19)$$

and $Q_n(h)$ is determined from

$$\int_1^h \frac{P_n(h) - P_n(\zeta)}{h - \zeta} d\zeta$$

Consequently, $P_n(h)$ has integral coefficients and $Q_n(h)$ has a common denominator lcm$\{1,\ldots,n\}$. For various h and various ratios m/n, various Padé-type approximations to ln h are obtained. We refer to Ref. 18 for various recurrence relations and applications to h = 2.

For imaginary quadratic h, this approach provides a rational approximation to inverse trigonometric functions. It is better to use different

Algebra for Diophantine Equations 31

forms of integrals in (2.7), with two distinct poles instead of one in (2.11). These integrals have the form

$$I = \int_0^1 \frac{[\zeta(\zeta - 1)]^{v_0} \prod_{k \geq 2} P_k(\zeta)^{v_k}}{(\zeta - z)^{n+1}[\zeta - (z - 1)]^{n+1}} d\zeta \qquad (2.20)$$

The integrals of the form (2.20) give the rational approximations to

$$\frac{1}{2z - 1} \ln\left[\left(1 - \frac{1}{2}\right)^2\right]$$

when $(2z - 1)^2$ is positive. When $(2z - 1)^2 = -\Delta$ is negative (i.e., when $z = 1/2 + it$), the integral (2.20) gives rational approximations to

$$\frac{1}{\sqrt{\Delta}} \arctan \frac{1}{\sqrt{\Delta}}$$

We elaborate on integrals of the type (2.20) in connection with rational approximations to $\pi/\sqrt{3}$ and π. One of the important formulas here is the following:

$$\int_0^1 \frac{dx}{[x(x - 1) - \zeta(\zeta - 1)]^{n+1}}$$

$$= \sum_{k=0}^{n-1} \frac{2^{k+1}(2n - 1)\cdots(2n - 2k + 1)}{n\cdots(n - k)} (-1)^{n+1}$$

$$\times (2\zeta - 1)^{-2k-2}[\zeta(\zeta - 1)]^{-n+k} + \binom{2n}{n}(-1)^n (2z - 1)^{-2n} \alpha \qquad (2.21)$$

Here $\alpha = 1/(2\zeta - 1) \ln[(\zeta - 1)^2/\zeta^2]$ for $\zeta \in \mathbb{R}$ and $\alpha = (4/\sqrt{\Delta}) \arctan(1/\sqrt{\Delta})$ if $\Delta = -(2\zeta - 1)^2 > 0$. The change of variable $x(x - 1) = X$ maps the interval of integration $[0,1]$ onto $[-1/4,0]$ (see this and similar transformations in Ref. 19). Consequently, instead of the form (2.20) of integrals, we consider the following integral representation:

$$I' = \int_{u-1/4}^{u} \frac{\prod_{\xi}(Z - \beta_\xi)^{v_\xi} dZ}{Z^{n+1}\sqrt{4Z + (2\zeta - 1)^2}} \qquad (2.22)$$

where $u = -\zeta(\zeta - 1)$, and $\beta_0 = u$, $\beta_1 = u - 1/4 = -1/4(2\zeta - 1)^2$.

As follows from (2.22), any integral of the form I' (even with different path of integration) is a linear combination

$$I' = P' \frac{4}{\sqrt{\Delta}} \arctan \frac{1}{\sqrt{\Delta}} + Q' \qquad (2.23)$$

[for $\Delta = -(2\zeta - 1)^2 = 4u - 1$], where P' and Q; are rational functions in u, polynomials in β_ξ, and have the rational number coefficients.

The derivation of asymptotics of $|I'|$ in (2.22) and $|P'|$, $|Q'|$ using the steepest descent method is identical to that of integrals (2.11) and (2.11') because the main term $R(Z)^n = \Pi_\xi (Z - \beta_\xi)^{v_\xi}/Z^n$ in the integrand of (2.22) has a form identical to that of (2.11). In particular, the statement of asymptotics in Proposition 2.4 holds for (2.22) with the natural substitution of $R(\zeta)$ by $R(Z)$. The asymptotics of (2.14) and (2.15) as $n \to \infty$ also hold if one replaces the interval of integration $[1,h]$ by $[u - 1/4, u]$ in (2.22).

The only difference between integrals of the form (2.11) and (2.22) is in their arithmetic asymptotics, particularly in the term corresponding to Δ_n of Proposition 2.3. For the sake of completeness we present an analog of Proposition 2.3 for integrals of the form (2.22) and P', Q' from (2.23).

PROPOSITION 2.6 Let ζ and β_ξ be algebraic numbers from the field K. Let S be the set of all nonarchimedean valuations v of K such that $v((2\zeta - 1)^2/4) \neq 0$, $v(\zeta(\zeta - 1)) \neq 0$, or $v(\beta_\xi) \neq 0$ for some ξ. We put $t_v = \min\{v((2\zeta - 1)^2/4), v(\zeta(\zeta - 1))\}$, $b_v = \max\{v((2\zeta - 1)^2), v(\zeta(\zeta - 1))\}$. For $d = \Sigma_\xi v_\xi$, we put $M = \max\{n, 2(d - n)\}$, $\Delta_n = \text{lcm}\{1,\ldots,n\}$. Then the numbers $\Delta_n P'$ and $\Delta_n Q'$ for P' and Q' from (2.23) are S-integral. Moreover, for any $v \in S$ we have

$$\min\{v(P'), v(Q')\} = \min\{\mu_1, \mu_2\}$$

where

$$\mu_1[= \mu_1(v)] = t_v(d - n) + \sum_{\xi, v(\beta_\xi) < t_v} v_\xi(v(\beta_\xi) - t_v)$$

$$\mu_2[= \mu_2(v)] = b_v(d - n) + \sum_{\xi, v(\alpha_\xi) < b_v} v_\xi(v(\beta_\xi) - b_v)$$

In particular, if P are prime ideals of K corresponding to $v \in S$, then

$$D = \Delta_n \prod_{P, v_p \in S} P^{-\min\{\mu_1(v_p), \mu_2(v_p)\}}$$

is the common denominator of $P', Q' \in K$ from (2.23).

Algebra for Diophantine Equations

Note a significant difference between Propositions 2.6 and 2.3. In the simplest case when $v_k = 0$ for $k \geq 2$ and $v_0 = v_1$ with $\alpha_0 = 1$, $\alpha_1 = h$ in (2.11), and $\beta_0 = u$, $\beta_1 = u - 1/4$ in (2.22), the contribution to the denominators P and Q in (2.12) from primes p such that $v_p(h) = 0$ is $\text{lcm}\{1,\ldots,M\} = e^{\psi(M)} \sim e^M$ for $M = \max\{2v_0 - n, n\}$. The contribution to the denominator of P', Q' in (2.23) from primes p such that $v_p(u) = v_p(u - 1/4) = 0$ is, on the other hand, $\text{lcm}\{1,\ldots,M\} \sim e^M$ for $M = \max\{4v_0 - 2n, n\}$. Also, the asymptotics of $1/n \log |I'|$ in (2.22) for $v_0 = v_1 = n$ is exactly twice that of $1/n \log |I|$ in (2.11) for $v_0 = v_1 = n$.

Consequently, one can improve the figure of merit given by Lemma 2.5 (ratio a/b) by taking v_0, v_1 in (2.22) to be less than n. An optimal value of $v_0 + v_1$ is around $3n$ for those ζ for which $(\zeta - 1)/\zeta$ is a unit [or $1/\sqrt{\Delta}$ arc tan $(1/\sqrt{\Delta})$ is commensurable with π]. The case of $\pi/\sqrt{3}$ is the most revealing. This corresponds to $\zeta = (1 + \sqrt{-3})/2$. Padé approximations of the form (2.23) corresponding to (2.22) for this ζ with $v_0 = v_1 = [3n/4]$, $v_k = 0$ for $k \geq 2$ (and n divisible by 4) were analyzed in detail in Section 5 of Ref. 18. In this case we presented three-term recurrences relating consecutive P' and Q' that have coefficients polynomial in n. The derivation of these and other properties had been achieved in the IBM SCRATCHPAD environment.

The sequences of dense approximations to $\pi/\sqrt{3}$ arising from (2.22) and (2.23) with $v_0 = v_1 = [3n/4]$, $v_k = 0$ for $k \geq 2$ give, according to Lemma 2.5 (and the arithmetic asymptotics of Proposition 2.6), the following exponent of rational approximations to $\pi/\sqrt{3}$:

$$\left| q \frac{\pi}{\sqrt{3}} - p \right| > |q|^{-4.8174417\ldots}$$

for rational integers p, q with $|q| \geq q_1$ (see Ref. 18). Slightly different choice of v_0 and v_1 can improve this exponent. For example, if one puts $v_0 = [0.8n]$, $v_1 = [0.7n]$ ($v_k = 0$ for $k \geq 2$), the following measure of rational approximation is achieved:

$$\left| q \frac{\pi}{\sqrt{3}} - p \right| > |q|^{-4.792613804\ldots}$$

for $|q| \geq q_2$. (As we noted before, the process of improvement of the exponent by means of adding terms to the integrand in (2.22) can be continued indefinitely. For example, taking $v_0 = [0.8005n]$, $v_1 = [0.6995n]$ improves the exponent to $-4.7926098\ldots$. Addition of two more zeros to the integrand improves this exponent further.)

Another interesting application of integrals (2.22) is to the study of π. Although several good measures of irrationality of π are known (see Refs. 7 and 18), all these measures are achieved by means of multidimensional integral representations. As a result, all these measures are based on approximation of π and its high powers (involving also π^2). Not a single system of dense approximations to π of the type demanded by Lemma 2.5 was known. Integrals of the form (2.22) with $\zeta = (1 + i)/2$ and $\Delta = 1$ provide such sequences of dense approximations to $\pi = 4 \arctan 1$. A simple search for appropriate integrands of the form (2.22) leads to the following simple class of integrals:

$$I' = \int_{1/4}^{1/2} \frac{(2Z - 1)^{v_0} (4Z - 1)^{v_1} (3Z - 1)^{v_2}}{Z^{n+1} \sqrt{4Z - 1}} dZ \qquad (2.24)$$

A choice of v_0, v_1, v_2 as $v_0 = [0.435n]$, $v_1 = [0.305n]$, $v_2 = [0.755n]$ leads to the dense sequence of approximations $D_n I' = P_n' \pi - Q_n'$ to π with integral P_n', Q_n' satisfying the condition of Lemma 2.5. This choice of parameters in (2.24) leads to an exponent in the measure of irrationality of π of $-55.9829\ldots$. This exponent is far from the best that we obtained, but a simple system of rational approximations Q_n'/P_n' to π is important for various problems in approximations to π. We note that from contiguous relations (2.6) for Pochhammer integrals it follows that sequences P_n' and Q_n' satisfy the same four-term linear recurrence with coefficients polynomial in n.

3. DIOPHANTINE APPROXIMATIONS TO VARIOUS NUMBERS

In this section we look at two classes of numbers: classical transcendence constants (numbers from Section 2) and algebraic numbers of the form $\sqrt[n]{\alpha}$, for which exponents of irrationality are better than those given by the Liouville theorem. In all cases we are using Padé-type approximations expressed by multidimensional integrals of integrands that are products of rational functions with rational exponents. As in Section 2 we use the WKB method to determine the asymptotics of the corresponding Padé-type approximants and the remainder functions (to satisfy Lemma 2.5 on dense sequences and its immediate generalizations [40]).

Often one can determine explicitly the linear recurrences connecting the consecutive Padé approximants, and thus have an asymptotics of Padé approximants and their remainder functions (as in Lemma 2.5) from the

Algebra for Diophantine Equations

asymptotic form of the corresponding linear recurrence. The case in point is an example of Section 2 on approximations to $\pi/\sqrt{3}$ that gives the exponent of irrationality $-4.8174417\ldots$. Let us present, following Ref. 18, expressions for the corresponding sequences P' and Q' in (2.23) and the three-term linear recurrence connecting consecutive approximants. Let us look at a dense sequence X_n/Y_n approximating $4\pi/3\sqrt{3}$, where the Y_n are rational integers and the X_n are rational numbers with denominators dividing $\mathrm{lcm}\{1,\ldots,4n\}$. The expression of Y_n is the following:

$$Y_n = \sum_{\substack{i_1=0 \\ i_1+i_2 \leq 4n}}^{3n} \sum_{i_2=0}^{3n} \binom{3n}{i_1}\binom{3n}{i_2}\binom{2(4n-i_1-i_2)}{4n-i_1-i_2}(-1)^{i_1+i_2} 4^{i_2} 3^{i_1} \qquad (3.1)$$

A study of the integral representations (2.22) shows the existence of a scalar three-term linear recurrence relation satisfied by each of the sequences X_n and Y_n that has coefficients polynomial in n:

$$\begin{aligned} A_2(n)Y_{n+2} + A_1(n)Y_{n+1} + A_0(n)Y_n &= 0 \\ A_2(n)X_{n+2} + A_1(n)X_{n+1} + A_0(n)X_n &= 0 \end{aligned} \qquad (3.2)$$

Here $A_0(n)$, $A_1(n)$, and $A_2(n)$ are polynomials in n of degree 9 with integer coefficients and are explicitly the following:

$$\begin{aligned} A_2(n) = &-2^3(4n+7)(4n+5)(4n+3)(4n+1)(2n+3)(n+2) \\ &(27{,}279n^3 + 52{,}164n^2 + 31{,}511n + 6046) \\ A_1(n) = &\;3(4n+3)(4n+1)(15{,}484{,}624{,}281n^7 + 12{,}251{,}806{,}648n^6 \\ &+ 401{,}859{,}218{,}160n^5 + 706{,}125{,}904{,}254n^4 + 715{,}282{,}318{,}379n^3 \\ &+ 415{,}975{,}459{,}648n^2 + 128{,}021{,}157{,}420n + 16{,}022{,}087{,}856) \\ A_0(n) = &\;2\cdot 3^3(6n+5)(6n+1)(3n+2)(3n+1)(2n+1)(n+1) \\ &(27{,}279n^3 + 134{,}001n^2 + 217{,}676n + 117{,}000) \end{aligned}$$

The initial conditions and the first few terms X_n and Y_n are the following:

$Y_0 = 1 \quad Y_1 = 1250 \quad Y_2 = 5{,}915{,}250 \quad Y_3 = 32{,}189{,}537{,}978,\ldots$

$X_0 = 0 \quad X_1 = 3023\ldots \quad \text{and} \quad X_2/Y_2 = 111{,}264{,}499/46{,}007{,}500,\ldots$

The limit quadratic equation that determines the asymptotics of Y_n and X_n follows from (3.2): $2^8 x^2 - 3^3 \cdot 59 \cdot 1069 x - 3^5 = 0$. This implies, according to Lemma 2.5, the following measure of irrationality of $\pi/\sqrt{3}$:

$$\left| q \frac{\pi}{\sqrt{3}} - p \right| > |q|^{-4.8174417\ldots}$$

for arbitrary rational integers p, q, with $|q| \geq q_1$.

The expression for Y_n can be derived by summing the inner sum in (3.1):

$$Y_n = \sum_{i=0}^{3n} \binom{3n}{i} \binom{-1/2 + 3n}{4n - i} 4^{4n-i} 3^i \tag{3.3}$$

Higher-order linear recurrences can be deduced for more complicated dense sequences of approximations from Section 2 [e.g., four-term linear recurrence for integrals of the form (2.24)].

Let us turn now to algebraic numbers. The Liouville theorem guarantees that for an algebraic number α of degree n there exists an *effective* constant c > 0 such that for arbitrary rational integers p, q : $|\alpha - p/q| > c|q|^{-n}$. The problem of improving the exponent n is of crucial importance in a variety of diophantine problems. Since Thue [54] there had been significant progress in this area, although the strongest results are ineffective and cannot be applied immediately to the effective solution of diophantine equations. All these improvements over the Liouville theorem have the following form:

$$\left| \alpha - \frac{p}{q} \right| > c(\alpha, \lambda) |q|^{-\lambda} \tag{3.4}$$

for rational integers p, q with $|p|, |q| \geq q_0(\alpha, \lambda)$. Here λ is any $\lambda > n/2 + 1$ (Thue theorem), $\lambda > \min_{s=1,\ldots,n-1} \{n/(s + 1) + s\}$ (Thue-Siegel theorem), $\lambda > \sqrt{2n}$ (Gelfond-Dyson theorem), and λ is any > 2 (Thus-Siegel-Roth theorem). However, the corresponding constants $c(\alpha, \lambda)$ and $q_0(\alpha, \lambda)$ are *ineffective*. This means that (possibly) finitely many counterexamples to the inequality (3.4) are allowed, and the sizes of these counterexamples are unknown (although their number can be explicitly bounded). All these results were proved using different ingenious variations of the Thue method, where the existence of one counterexample to (3.4) with large $|q|$ implies a good (effective) measure of irrationality of α. The Roth method and later the Schmidt method [13], which deals with simultaneous approximation to several algebraic numbers, uses, as an initial step, a large number of very good rational approximations. Consequently, these methods have a chance to turn effective if numbers α to which they are applied possess effectively determined "good" rational approximations of relatively large sizes. The possibility of using this approach to gain partial effectivization was first discussed by Thue. Mahler [55] made the first explicit statement about such effective results (in archimedean and p-adic metrics) in the context of the Thue-Siegel theorem. Later, Gelfond [56] made a similar assertion in the context of the Dyson-Gelfond theorem, which was elaborated on by Hyyro. Recently, Bombieri [58] significantly generalized and improved this approach (in the context of bounds similar to the Dyson-Gelfond theorem).

The only known *general* effective improvement on the Liouville theorem had been achieved by <u>Baker's method</u> on lower bounds of linear forms in logarithms of algebraic numbers [3]. This method allows us to find an effective $\lambda < n$ in (3.4) such that $c(\alpha,\lambda)$ and $q_0(\alpha,\lambda)$ can be effectively determined. These results, although weak in comparison with the corresponding ineffective statements, immediately imply effective bounds on solutions of a variety of diophantine equations, including the Thue equations. Unfortunately, the difference $n - \lambda$ (and the constant c) is very small [i.e., $1/(n - \lambda)$ is of the order $\exp\{0(D_{\mathbb{Q}(\alpha)/\mathbb{Q}})^{1/2+\varepsilon}\}$] and numerical constants involved in the expressions of λ, c, and q_0 are too large to allow (3.4) to be easily applicable. These limitations can be attributed to the "ineffectiveness" of the method of *proofs* in the Baker's technique, where certain auxiliary approximating functions are not constructed but rather proved to exist via a certain version of Dirichlet's box principle. In contrast, whenever effective construction of approximating functions is known, a good effective bound can be expected. As Thue noted, Padé approximations to $\sqrt[n]{1-x}$ at $x = 0$ are expressed in terms of Jacobi polynomials, and thus some effective results can be proved for algebraic numbers $\sqrt[n]{b/a}$ whenever b/a is close to 1 (or, whenever $\sqrt[n]{b/a}$ has a very good known rational approximation). Mahler generalized these considerations using Hermite's effective families of Padé approximations to general binomial functions $(1 - x)^{\omega_1}, \ldots, (1 - x)^{\omega_m}$. We refer the reader to our paper (Ref. 40) for all relevant information and references. Hypergeometric polynomials associated with Padé approximants to $\sqrt[n]{1-x}$ were a subject of particular attention [1,21,52,57] (see additional references in Ref. 40). We wish to single out the work of Baker [57], where the first strong results for numbers such as $\sqrt[3]{2}$ and $\sqrt[3]{17}$ had been obtained. In Ref. 40 we determined the asymptotics of the remainder function and Padé approximants, and we found that the p-adic convergence of the remainder function is determined by a peculiar law of distribution of primes in short arithmetic progressions. This allowed us to prove in Ref. 40 some strong results for measures of irrationality of large classes of algebraic numbers (particularly, for general cubic irrationalities). In this section we report briefly on the extension of the work of Ref. 40. We are able to determine explicitly by the distribution of primes in the denominators of coefficients of Padé approximants to binomial functions $(1 - x)^{\omega_i}$: $i = 1$, ..., m. We determine the asymptotics of these denominators (in the diagonal case and for large weights) in terms of the values of $\psi(z) = d \log \Gamma(z)/dz$. This allows us to prove the effectivization of Thue and Thue-Siegel theorems for an important class of algebraic numbers of large degrees. A particularly important class of algebraic numbers is given by all numbers from the Kummer fields $\mathbb{Q}(\sqrt[n]{\alpha})$ for a fixed (algebraic) α and large n. For *numbers* $\sqrt[n]{\alpha}$ themselves, an effective improvement over the Thue-Siegel theorem holds according to Baker [3]. Equation (3.4) is true with $\lambda = c_1(\alpha) \log n$, where $c_1(\alpha)$ is an effective constant (but the same remark concerning the sizes of constants applies here). Unfortunately, Baker's results do not give the same improvement over the Liouville theorem for all numbers from $\mathbb{Q}(\sqrt[n]{\alpha})$. Similarly, ineffective methods of proof (see Ref. 58) gave nothing for fields $\mathbb{Q}(\sqrt[n]{\alpha})$ with fixed α and large n. Padé approximation methods apparently are efficient here. We show that for every algebraic α there exists a computable constant $c_2(\alpha)$ such that we have an effective version of the Thue theorem for *all* members of the field $\mathbb{Q}(\sqrt[n]{\alpha})$ whenever $n \geq c_2(\alpha)$; for example, we have an effective improvement over the Liouville theorem for all numbers $\sqrt[n]{2}$ for $n \geq 23$. Another application involves lower bounds for two logarithms in algebraic numbers. The corresponding bounds had been derived for the first time by Gelfond in his solution of Hilbert's seventh problem. As it turns out, lower bounds for linear forms with integer rational coefficients in two logarithms can be deduced from Padé approximants to binomial functions. This allows for a certain improvement in the effective bounds for solutions of diophantine equations, when two logarithms are sufficient. Among diophantine equations from this class we can mention $f_3(x,y) = 1$, where f is a cubic integral form of negative discriminant. Also, p-adic diophantine approximations can be treated the same way in all cases studied below.

The key to the analysis of linear forms in two (and, sometimes, even in three) logarithms lies in replacement of the approximation problem $|\alpha^j \beta^n - 1|$ for large integers j and n by $|(\alpha^k \beta^\ell)^{m/s} - \alpha^{m_0/s} \beta^{m_1/s} 1|$ for large m and s (of logarithmic sizes comparable with that of j and n), and relatively small k and ℓ but such that $\alpha^k \beta^\ell$ is very close to 1 [or, for proper branches of $\log(\cdot)$, k/ℓ is a very good rational approximation to

$-\log \beta/\log \alpha$]. The focus switches to the approximations of $\alpha^{m/s}$ for α (close to 1) and large m and s. We use for this problem Padé approximants to binomial functions studied in Ref. 40. The remainder function in this Padé approximation problem to $(1 - x)^{\omega_i}$: $i = 1, \ldots, m$ at $x = 0$ with weights n_i: $i = 1, \ldots, m$ is expressed as a multidimensional integral (derived by Mahler)

$$R(n_1,\ldots,n_m;x) = \int_0^x dt_1 \int_0^{t_1} dt_2 \cdots \int_0^{t_{m-2}} dt_{m-1} (x - t_1)^{n_1-1}$$
$$\times (t_1 - t_2)^{n_2-1} \cdots t_{m-1}^{n_m-1} (1 - x)^{\omega_1} (1 - t_1)^{\omega_2-\omega_1-n_1} \cdots$$
$$\times (1 - t_{m-1})^{\omega_m-\omega_{m-1}-n_{m-1}} \quad (3.5)$$

This integral is of the form studied in Section 2, and the coefficients of the Padé approximants in this problem are expressed as products of binomial coefficients. For example, for $n_i = r + 1$ these coefficients have the form

$$\frac{(r!)^{m-1}}{\Pi_{j=1,j\neq i}^m [(\omega_j - \omega_i + r - h - 1)\cdots(\omega_j - \omega_i)\cdots(\omega_j - \omega_i - h)]} \binom{r}{h} \quad (3.6)$$

for $i = 1, \ldots, m$ and $h = 0,1, \ldots, r$. The common denominator of the coefficients of Padé approximants to $(1 - x)^{\omega_i}$ is determined as (the smallest) rational number $B_r(\omega_1,\ldots,\omega_m)$ such that $B_r(\omega_1,\ldots,\omega_m)$ times (3.6) is an integer for all $i = 1, \ldots, m$ and $h = 0, \ldots, r$. The archimedean asymptotics of (3.5) and the corresponding Padé approximants were completely determined in Section 2 of Ref. 40. The arithmetic part of the asymptotic is determined for given rational numbers $\omega_1, \ldots, \omega_m$ with denominator n as the following quantity $(\text{Chr})_n^m = \lim_{r\to\infty} (1/r) \log B_r(\omega_1,\ldots,\omega_m)$. We study the primes p entering $B_r(\omega_1,\ldots,\omega_m)$ for $\omega_i = k_i/n$: $i = 1, \ldots, m$. Methods of Sections 4-6 of Ref. 40 (similar methods are used in Section 7 of this book) allow us to determine the intervals where p lies according to its congruence properties mod nm. Using computer algebra systems we developed a program that determines $(\text{Chr})_n^m$ explicitly in terms of linear forms of logarithms of algebraic numbers. Then choosing appropriate k_i and m, we were able to determine the best possible exponents in the effective improvement over the Liouville theorem.

Table 1 Exponents λ in (3.4) for All Numbers from Cubic Field $\mathbb{Q}(\sqrt[3]{D})$ and Effective c, q_0

D	λ
2	2.428...
3	2.690...
5	2.998...
6	2.320...
7	2.725...
10	2.413...
11	2.997...
12	2.907...
13	2.824...

In Table 1 we list the exponents λ in (3.4) for cubic irrationalities that improve on the corresponding table in Ref. 40. (For some of the numbers, approximations to $\varepsilon^{j/n}$ for fundamental units of cubic fields were considered.)

Sometimes one can prove more impressive-looking results. For example, if $n \geq 89$, every algebraic number from the field $\mathbb{Q}(\sqrt[n]{2})$ satisfies the effective version of the Thue theorem:

$$\left|\alpha - \frac{p}{q}\right| > |q|^{-(n/2)-1}$$

for $|q| \geq q_0(\alpha)$ and effective constant $q_0(\alpha)$ depending on $\alpha \in \mathbb{Q}(\sqrt[n]{2})$.

4. MODULAR EQUATIONS AND APPROXIMATIONS TO π

The title of this section is the name of one of the famous Ramanujan papers [25] published in 1914 summarizing much of Ramanujan's original work in India. As the title suggests, there are two parts to this work of Ramanujan. Whereas the first part, dealing with modular equations, has been the subject of much later work (and much early work of which Ramanujan was unaware), the second part and its relation with the first were left pretty much in the state developed by Ramanujan. We refer to Hardy's book [26] for further references to work on modular equations, including important papers of Watson. In this section we describe Ramanujan-like convergent hypergeometric series representing π and discuss some of their applications to diophantine approximations of numbers connected with π. Ramanujan-like

Algebra for Diophantine Equations

series in general are a part of a very general class of relations between periods and quasi-periods of abelian varieties with complex multiplication and solutions to Picard-Fuchs equations evaluated at moduli points corresponding to these varieties. (We deal in later publications with the general situation; here we discuss only elliptic curves.)

One of the most obvious relations between modular equations and approximations to π stems from a crucial observation that $j(\omega)$ is an algebraic integer whenever $\omega \in \mathbb{Q}(\sqrt{-d})$ for $d > 0$, and $j(\omega)$ has the degree $h(-d)$ whenever $\{1,\omega\}$ is a fractional ideal of $\mathbb{Q}(\sqrt{-d})$. Here $j(\omega)$ is an absolute (modular) invariant of $\omega \in H$ (i.e., Im $\omega > 0$), and $h(-d)$ is the class number of the imaginary quadratic field $\mathbb{Q}(\sqrt{-d})$. In particular, if an imaginary quadratic field $\mathbb{Q}(\sqrt{-d})$ has class number 1, then for its generator ω, $j(\omega)$ is a rational integer. On the other hand, $j(\omega)$ is represented by a convergent q-series. We normalize $j(\omega)$ later and now we consider one of its representations as

$$j(\omega) = q^{-1} + 744 + 196{,}884q + 21{,}493{,}760q^2 + \cdots$$

where $q = e^{2\pi i \omega}$. Then for $\omega = (1 + \sqrt{-d})/2$ for $d \equiv 3(8)$, $q^{-1} = -e^{\pi\sqrt{d}}$ appears to be very close to an algebraic integer of degree $h(-d)$. Taking $-d$ as one of the one-class discriminants, we obtain various interesting approximate results. For example, for $d = 163$ (the largest such discriminant) we have

$$e^{\pi\sqrt{163}} = 262{,}537{,}412{,}640{,}768{,}743.9999999999992\ldots \quad (4.1)$$

(very close to an integer, indeed).

It appears that those who begin the work on modular equations were all interested in approximate relations such as (4.1). Among them are Hermite [27] (1858), Kronecker (1863), and Smith (1865). Hermite [27] gives an approximation similar to (4.1) for $d = 43$: $e^{\pi\sqrt{43}} = 884{,}736{,}474.999777\ldots$. Ramanujan rediscovers these approximations in Section 4 of Ref. 25. About a century later, approximations such as (4.1) became the basis of one of the solutions of class 1 problems, and the basis to the solution of class 2 problems. By that time different forms of approximate modular equations had been discovered by Stark, who deduced them from the Kronecker limit formula [28]. In these formulas it is not log $j(\omega)$, which is approximated by $\pi\sqrt{d}$, but rather a linear form in two (or three) logarithms. For example, in one of these formulas,

$$L(1,\chi)L(1,\chi\chi') = \frac{\pi^2}{6} \prod_{p|k} \left(1 - \frac{1}{p^2}\right) \sum_f \frac{\chi(a)}{a} \qquad (4.2)$$

is represented by a fast convergent series $\Sigma_f \Sigma_{r=-\infty}^{\infty} A_r e^{\pi i r b/ka}$, where $\chi(n) = (k/n)$, $\chi'(n) = (-d/n)$, $f = ax^2 + bxy + cy^2$ runs through a complete set of inequivalent quadratic forms with the discriminant $-d$, and $k > 0$. Since $L(1,\chi) = 2h(k) \log \varepsilon_{\sqrt{k}}/\sqrt{k}$, $L(1,\chi\chi') = h(-kd)\pi/\sqrt{kd}$ by Dirichlet formulas, one deduces very sharp upper bounds on linear forms in two logarithms in algebraic integers: $|h_1 \log \varepsilon\sqrt{k_1} + b\pi \sqrt{d}|$ or $|h_1 \log \varepsilon\sqrt{k_1} + h_2 \log \varepsilon\sqrt{k_2}|$ are bounded by $e^{-\pi O(\sqrt{d})}$, whenever $-d$ is a class one discriminant. These inequalities were generalized to linear forms in three logarithms in the class 2 case (see Ref. 3), and recently new special relations of this kind for several logarithms were proposed by Shanks [29]. What is remarkable about Ramanujan's approximations of Section 8 of Ref. 25 and their generalizations is the ability to represent *each* of the terms in the linear form by a rapidly convergent algebraic series. For example, it is not only correct that $\log j(\omega)$ [or $\log(-j(\omega) + 744)$] is very close to $\pi\sqrt{d}$ as in (4.1), but, in fact, both of these numbers can be represented as rapidly convergent algebraic power series in $j(\omega)^{-1}$! The same is true for each of the terms in the linear forms in logarithms arising from the Kronecker limit formula (4.2). To understand better the essence of these representations, we need a brief remainder of the theory of complex multiplication. We refer the reader to Weil [30] for an exposition of the theory of elliptic modular functions.

Instead of more general notations of Weil [30], we use more traditional notations of Eisenstein series E_n as

$$E_2(\tau) = 1 - 24 \sum_{n=1}^{\infty} \sigma(n) q^n$$

$$E_4(\tau) = 1 + 240 \sum_{n=1}^{\infty} \sigma_3(n) q^n$$

$$E_6(\tau) = 1 - 504 \sum_{n=1}^{\infty} \sigma_5(n) q^n$$

$q = e^{2\pi i \tau}$.

The standard theory of complex multiplication states that for an arbitrary elliptic curve E with complex multiplication by $\sqrt{-d}$ and with periods ω_1, ω_2: $\tau = \omega_1/\omega_2 \in H$ all ratios $E_{2n}(\tau) : (\omega_2/2\pi i)^{2n}$ (for $n > 1$)

Algebra for Diophantine Equations

are algebraic integers. [This is essentially equivalent to the statement of algebraicity of $j(\tau)$ for $\tau = \mathbb{Q}(\sqrt{-d})$ made at the beginning of the section.] Ramanujan in Section 8 of Ref. 25 makes an additional algebraicity statement. Its rigorous formulation can be found in Section 3 of Chapter 6 in Ref. 30. One can assume that it belongs to Kronecker. According to this statement, algebraicity result also holds for $E_2(\tau)$ if an additional nonholomorphic term is introduced.

LEMMA 4.1 If $\tau \in \mathbb{Q}(\sqrt{-d})$, then the nonholomorphic series $[E_4(\tau)/E_6(\tau)] \times [E_2(\tau) - 3/(\pi \text{ Im}(\tau))]$ has an algebraic value [which belongs to a Hilbert class field $\mathbb{Q}(\sqrt{-d},j(\tau))$].

This is an "additional" period relation for elliptic curves found by Ramanujan. It plainly means the following: Between periods ω_1, ω_2 and quasi-periods η_1, η_2 of an elliptic curve E there is always a Legendre relation

$$\omega_2 \eta_1 - \omega_1 \eta_2 = 2\pi i \qquad (4.3)$$

and in the case of curves with complex multiplication, there is one more linear relation: $\omega_2/\omega_1 \in \mathbb{Q}(\sqrt{-d})$; in fact, in the complex multiplication case, a third relation occurs.

There exists a quadratic relation between ω_2 and η_2. To express this relation in a convenient form, we follow the notation of Ref. 31 and put (compare with Lemma 4.1)

$$s_2(\tau) = \frac{E_4(\tau)}{E_6(\tau)} \left[E_2(\tau) - \frac{3}{\pi \text{ Im}(\tau)} \right]$$

Let an elliptic curve E with periods and quasi-periods ω_1, ω_2, η_1, η_2 as above have a complex multiplication by $\sqrt{-d}$, where

$$\tau = \frac{\omega_1}{\omega_2} = \frac{1 + i\sqrt{d}}{2}$$

Let g_2, g_3 be the Weierstrass invariants of E that are algebraic integers from $\mathbb{Q}(\sqrt{-d},j(\tau))$. Then we arrive at Ramanujan's identity,

$$\omega_2 \eta_2 \sqrt{-d} + \omega_2^2 \left[\sqrt{-d} \, \frac{3g_3}{2g_2} \, s_2(\tau) \right] = 2\pi i \qquad (4.4)$$

By itself, (4.4) is not yet an approximation to π and one has to express ω_2, η_2 as appropriate convergent series. For this we apply a hypergeometric function representation of functions inverse to elliptic modular ones à la Schwarz and Fricke-Klein. It is appropriate to take $z = 12^3/j(\tau)$ as an elliptic function that we are inverting. We refer the reader to Fricke [32], where the corresponding theory of $_2F_1(1/12, 5/12; 1; z)$ hypergeometric functions is developed [in complete analogy with Legendre's $_2F_1(1/2, 1/2; 1; z)$-functions for $z = k^2(\tau)$ for modular functions of level 2]. We just remark on a few of the relations between periods and quasi-periods and their dependence on the modular invariant. If ω_k and η_k are normalized by means of the discriminant Δ of E as follows:

$$\Omega_k = \omega_k \Delta^{1/12} \qquad H_k = \eta_k \Delta^{-1/12}$$

we obtain the following differential relation:

$$\eta_k = -2\sqrt{3} \, J^{2/3}(J-1)^{1/2} \frac{\partial \Omega_k}{\partial J} \Delta^{1/12}$$

$$\omega_k = \Omega_k \Delta^{-1/12} \tag{4.5}$$

Here $J = J(\tau)$ is chosen in Fricke's normalization as

$$J(\tau) = \frac{j(\tau)}{12^3}$$

The function $\Omega_2 = \Omega_2(\tau)$ is given by Fricke's hypergeometric function

$$F(z) \stackrel{\text{def}}{=} {}_2F_1\left(\frac{1}{12}, \frac{5}{12}; 1; z\right)$$

as follows:

$$\Omega_2 = 2\pi j^{-1/12} F\left(\frac{12^3}{j}\right) \tag{4.6}$$

Combining identity (4.4) with (4.5) and (4.6), we obtain the following explicit generic expression of all Ramanujan-like rapidly convergent series representing π or, rather, $1/\pi$:

$$F\left(\frac{12^3}{j}\right)^2 \frac{1}{12}[1 - s_2(\tau)] + F\left(\frac{12^3}{j}\right) F'_z\left(\frac{12^3}{j}\right) \frac{12^3}{j} = \frac{j^{1/2}}{(j - 12^3)^{1/2} 2\pi\sqrt{d}} \tag{4.7}$$

Here in (4.7), $j = j(\tau)$ and $\tau = (-1 + \sqrt{-d})/2$ for $d > 0$, $d \equiv 3(4)$, and as above,

$$F(z) = {}_2F_1\left(\frac{1}{12}, \frac{5}{12}; 1; z\right) = \sum_{n=0}^{\infty} \frac{(1/12)_n (5/12)_n}{n!^2} z^n$$

Algebra for Diophantine Equations

Several important points connected with (4.7) should be made. First, according to Lemma 4.1, $s_2(\tau)$ is an algebraic number from a real subfield of a Hilbert field $\mathbb{Q}(\sqrt{-d}, j(\tau))$. Second, according to the Weber-Heegner result (see Stark [28]), $j^{1/3}$ and $(j - 12^3)^{1/2}/\sqrt{-d}$ are (real) algebraic integers of degree $h(-d)$. It clearly means that for a class 1 discriminant $-d$, all coefficients on the left side of (4.7) are rational numbers, while on the right side of (4.7) we have a rational multiplier of $(-j)^{1/6}/\pi$, where $(-j)^{1/6}$ is a quadratic irrationality. The relation (4.7) itself is not very convenient because we are dealing with squares of the function $F(z)$. It is here where Ramanujan's knowledge of hypergeometric function identities proved its importance. Ramanujan knew several instances when squares of hypergeometric functions are $_3F_2$ hypergeometric functions. In our case we need only a corollary of Clausen's theorem:

$$_2F_1\left(\frac{1}{12}, \frac{5}{12}; 1; z\right)^2 = {}_3F_2\left(\frac{1}{6}, \frac{5}{6}, \frac{1}{2}; 1, 1; z\right) \tag{4.8}$$

Equation (4.8) immediately allows us to represent (4.7) in Ramanujan form. We put

$$R(z) \stackrel{\text{def}}{=} {}_3F_2\left(\frac{1}{6}, \frac{5}{6}, \frac{1}{2}; 1, 1; z\right) = \sum_{n=0}^{\infty} \frac{(6n)!}{(2n)! n!^3} 12^{-3n} z^n$$

Thus we have

$$\sum_{n=0}^{\infty} \left\{\frac{1}{6}[1 - s_2(\tau)] + n\right\} \frac{(6n)!}{(3n)! n!^3} \frac{1}{[j(\tau)]^n} = \frac{(-j)^{1/2}}{\pi} \frac{1}{[d(12^3 - j)]^{1/2}} \tag{4.9}$$

Expression (4.9) can be used whenever $\tau = (1 + \sqrt{-d})/2$. In this case the second factor in the right side of (4.9) is a rational integer. For example, for $d = 163$ we deduce from the known expression of $j((1 + \sqrt{-163})/2) = (-640,320)^3$ that [see (4.1)]

$$\sum_{n=0}^{\infty} (c_1 + n) \frac{(6n)!}{(3n)! n!^3} \frac{(-1)^n}{(640,320)^{3n}} = \frac{1}{\pi} \frac{(640,320)^{3/2}}{163 \cdot 8 \cdot 27 \cdot 7 \cdot 11 \cdot 19 \cdot 127} \tag{4.10}$$

Here c_1 is a rational number, determined from Lemma 4.1 as 13,591,409/545,140,134. Identity (4.10) is the fastest convergent series with rational entries among all Ramanujan-type series representations of multiples of $1/\pi$. One notices, however, that the Ramanujan series in Sections 13 and 14 of Ref. 25 do not have the structure of (4.9) and (4.10). This is connected with different choices by Ramanujan of hypergeometric functions used to invert elliptic modular functions, particularly with his interest in the Legendre function $_2F_1(1/2,1/2;1;z)$. The other three functions that Ramanujan chooses to represent periods of elliptic curves are $K_1(z) = {}_2F_1(1/4,3/4;1;z)$, $K_2(z) = {}_2F_1(1/3,2/3;1;z)$, and $K_3(z) = {}_2F_1(1/6,5/6;1;z)$. [The last function, $K_3(z)$, is easily expressed in terms of $F(z)$ above.] All these functions are related to $F(z)$ because the corresponding automorphic functions are algebraically dependent on $j(\tau)$ [or on $j(\alpha\tau)$ for various α]. The particular choice of $_2F_1$-function is explained by the need to have an analog of the Clausen formula for the square of the $_2F_1$-function as a $_3F_2$-function. Ramanujan obviously knew all such possible cases and among his favorite $_3F_2$-functions was $_3F_2(1/2,1/2,1/2;1,1;z)$. See equation (25) of Ref. 25, where Ramanujan writes: $(2K/\pi)^2 = 1 + (1/2)^3(2kk')^2 + (1.3/2.4)^3(2kk')^4 + \cdots$. Among various relations between this function and $R(z)$ we can mention only two: $_3F_2(1/2,1/2,1/2;1,1;z) = (1-z)^{-1/2} {}_3F_2(1/6,1/2,5/6;1,1;w)$ for $w = -27z/[4(1-z)^3]$, and $(1-z)^{-1/2} {}_3F_2(1/6,1/2,5/6;1,1;z) = {}_2F_1(1/12,5/12;1;z) {}_2F_1(11/12,7/12;1;z)$.

A very simple observation allows one to deduce from (4.9) a variety of similar formulas, including the Ramanujan equations (28)-(44) in Ref. 25. This observation is based on the existence of algebraic relations between automorphic functions for commensurable subgroups. Namely, if $\varphi_1(\tau)$ and $\varphi_2(\tau)$ are two automorphic functions with respect to the same fuchsian groups [which is a subgroup of $\Gamma(1) = SL_2(\mathbb{Z})$], then $\varphi_1(\tau)$ and $\varphi_2(\tau)$ are algebraically dependent, and in particular, $\varphi_1(\tau)$ is algebraically dependent on $j(\tau)$. Consequently, a power series $F(12^3/j(\tau))$ can be represented in terms of power series of $\varphi_1(\tau)$. Whenever $\varphi_1(\tau)$ is automorphic with respect to one of the triangle subgroups, one can replace $F(z)$ in (4.7) [or (4.9)] by an appropriate hypergeometric function. This observation is important because it allows us to express constants other than $1/\pi$ as values of convergent generalized hypergeometric series. This is true in particular for $\log \varepsilon_{\sqrt{D}}$ for a unit $\varepsilon_{\sqrt{D}}$ of a real quadratic field $\mathbb{Q}(\sqrt{D})$. This number can be represented as a value of generalized hypergeometric functions if we use the Kronecker limit formula expressing the logarithm in terms of products of values of Dedekind's Δ-function (*Jugendtraum*) (see Ref. 30). Such an expression of $\log \varepsilon_{\sqrt{D}}$ in terms of power series in $12^3/j(\tau)$ for $\tau = (1 + \sqrt{-d})/2$ depends, unfortunately, on D, because D is related to the level of the appropriate modular form. For $D = 5$ the necessary explicit formulas can be found in Siegel [33].

Algebra for Diophantine Equations

The importance of all these formulas for diophantine approximations is explained by the fact that the number π (or $j^{1/2}/\pi$ or $\ln \varepsilon_{\sqrt{D}}$) is represented as a value of the generalized hypergeometric function at a rational point very close to the origin. Thus, instead of interpreting π as a value of the logarithmic function, and using Padé-type approximations far from the point where we approximate, as we were doing in Sections 2 and 3, we can use ordinary Padé approximations to generalized hypergeometric functions. One group of such Padé approximations (of the second kind), provided in Ref. 9 is presented in detail in Section 5.

5. PADÉ APPROXIMATIONS TO GENERALIZED HYPERGEOMETRIC FUNCTIONS OF TYPE II

In this section we specialize and generalize the construction of Ref. 9 of Padé approximants of type II to generalized hypergeometric functions. The case of basic generalized hypergeometric functions is also considered. These Padé approximations are applied to Ramanujan-like representation of π and other logarithms of algebraic numbers of Section 4.

Let us recall Mahler's definitions [34] of Padé approximants of types I and II to the system $f_1(x), \ldots, f_m(x)$ of functions given by a formal power series at $x = 0$.

DEFINITION 5.1a For m functions $f_1(x), \ldots, f_m(x)$ given by a formal power series at $x = 0$ and m nonnegative integers $n_1, \ldots, n_m$, we consider m polynomials $A_1(x), \ldots, A_m(x)$ of degrees at most $n_1, \ldots, n_m$, respectively, such that the function

$$R(x) = A_1(x)f_1(x) + \cdots + A_m(x)f_m(x)$$

has a zero at $x = 0$ of the order of at least

$$\sum_{i=1}^{m} (n_i + 1) - 1$$

The polynomials $A_i(x)$ are called Padé approximants of type I and are denoted by $A_i(x|n_1,\ldots,n_m): i = 1, \ldots, m$. The function $R(x)$ is called the remainder function and is denoted by $R(x|n_1,\ldots,n_m)$.

DEFINITION 5.1b Let $f_1(x), \ldots, f_m(x)$ be formal power series and $n_1, \ldots, n_m$ be nonnegative integers. We say that the system of polynomials $(A_1(x), \ldots, A_m(x))$ is the system of Padé approximants of type II to the system of

functions $f_1(x), \ldots, f_m(x)$ with weights $n_1, \ldots, n_m$, if the following conditions are satisfied:
1. The polynomials $A_1(x), \ldots, A_m(x)$ are not all zero.
2. Deg $A_i(x) \leq \sum_{j=1, j\neq i}^{m} n_j = \sigma - n_i$ for $\sigma = n_1 + \cdots + n_m$, $i = 1, \ldots, m$.
3. The order of zero of the function $A_k(x) f_\ell(x) - A_\ell(x) f_k(x)$ at $x = 0$ is at least $\sum_{i=1}^{m} n_i + 1 = \sigma + 1$; $k, \ell = 1, \ldots, m$.

We denote Padé approximants of type II with weights $n_1, \ldots, n_m$ as follows: $A_i(x) = A_i(x|n_1, \ldots, n_m)$, $i = 1, \ldots, m$.

Mahler [34] proved an important relation between Padé approximants of types I and II. Mahler's relation can be represented in matrix form. We denote

$$A(x|n_1, \ldots, n_m) = (A_i(x|n_1 + \delta_{j1}, \ldots, n_m + \delta_{jm}))_{i,j=1}^{m}$$

$$\underline{A}(x|n_1, \ldots, n_m) = (A_i(x|n_1 - \delta_{j1}, \ldots, n_m - \delta_{jm}))_{i,j=1}^{m}$$

Then for some constants $c_1, \ldots, c_m$

$$A(x|n_1, \ldots, n_m) \underline{A}(x|n_1, \ldots, n_m)^t = \begin{bmatrix} c_1 x^\sigma & & 0 \\ & \ddots & \\ 0 & & c_m x^\sigma \end{bmatrix}$$

for $\sigma = n_1 + \cdots + n_m$. Padé approximants are of most interest in the diagonal case (equal weights $n_i = n$) or the near-diagonal case ($|n_i - n_j| \leq 1$).

The class of generalized hypergeometric functions we consider can be described using differential operators in terms of $\delta_x = x(d/dx)$. In general, the generalized hypergeometric function $f(x)$ is defined as a formal power series at $x = 0$ which is a solution of the linear differential equation

$$[Q(\delta_x) - xP(\delta_x)] f(x) = 0 \tag{5.1}$$

for two polynomials $P(x)$ and $Q(x)$. Hence $f(x) = \sum_{n=0}^{\infty} c_n x^n$, where $c_{n+1} = c_n [P(n)/Q(n+1)]$: $n = 0, 1, 2, \ldots$. Generalized hypergeometric functions are presented as functions of zeros of $P(x)$ and $Q(x)$. The standard representation of generalized hypergeometric functions involves the ratio of Γ-factors. We denote $(a)_0 = 1$, $(a)_n = a(a+1)\cdots(a+n-1)$, so that $(a)_\nu = \Gamma(a+\nu)/\Gamma(a)$. Then the ${}_pF_q$-hypergeometric function $f(x)$ is

Algebra for Diophantine Equations

$$f(x) = {}_pF_q\left(x \left|\begin{array}{c}a_1,\ldots,a_p\\b_1,\ldots,b_q\end{array}\right.\right) \stackrel{\text{def}}{=} \sum_{n=0}^{\infty} \frac{(a_1)_n \cdots (a_p)_n}{(b_1)_n \cdots (b_q)_n} \frac{x^n}{n!} \tag{5.2}$$

Then the $f(x)$ from (5.2) satisfies equation (5.1) with the following choice of polynomials $P(x)$ and $Q(x)$:

$$Q(x) = x(x + b_1 - 1)\cdots(x + b_q - 1)$$
$$P(x) = (x + a_1)\cdots(x + a_p) \tag{5.3}$$

The functions

$${}_pF_q\left(x \left|\begin{array}{c}a_1 + k_1,\ldots,a_p + k_p\\b_1 + \ell_1,\ldots,b_q + \ell_q\end{array}\right.\right)$$

for the integers k_i, ℓ_j are called contiguous functions to

$${}_pF_q\left(x \left|\begin{array}{c}a_1,\ldots,a_p\\b_1,\ldots,b_q\end{array}\right.\right)$$

In general, the module over $\mathbb{C}(x)$, generated by 1 and functions contiguous to $f(x)$ in (5.2), has a dimension of at most $\max\{p, q + 1\} + 1$ and is closed under differentiation. We will now study the system of Padé approximations of type II to the following system of functions: 1 and $\delta_x^j f(x)$: $j = 0, \ldots, \max\{p, q + 1\} - 1$, for a generalized hypergeometric function $f(x)$. Here $\delta_x^0 f = f$, $\delta_x f = xf_x$, and so on.

In order to present our results on Padé approximants of type II, we first reformulate and reprove the main result from Ref. 9:

THEOREM 5.2 [9] Let $f(x) = \sum_{n=0}^{\infty} c_n x^n$ be a regular solution of the generalized hypergeometric equation (5.1), and let $d_Q = \max\{\deg(P), \deg(Q)\}$. For two parameters n and b we put $r = [(n - 1)/d_Q] + n + b$. Let us define the polynomial $A(x) = A(x,n,b)$:

$$A(x) = \sum_{i=0}^{n} x^{n-i} g_{n-i} \frac{1}{c_{i+r-n}}$$

where $g_{n-i} = (-1)^i \binom{n}{i} P(i + b)\cdots P(i + r - n - 1)$: $0 \leq i \leq n$.

For every $k = 0,\ldots,(n - 1) - [(n - 1)/d_Q]d_Q$, we introduce the sequence $E_s^{(k)}$: $s = 0, 1, \ldots$ as follows:

$$E_s^{(k)} \stackrel{\text{def}}{=} \sum_{i=0}^{n} (i + s - n)^k g_{n-i} c_{i+s-n} \frac{1}{c_{i+r-n}}$$

Then for the following polynomial $B_k(x)$ of degree $n + b - 1$: $B_k(x) = \sum_{s=0}^{n+b-1} x^s E_s(k)$, we have the identity

$$A(x) \delta_x^k f(x) - B_k(x) = x^{r+1} \sum_{s=r+1}^{\infty} x^{s-r-1} E_s(k)$$

for all $k = 0, \ldots, (n - 1) - [(n - 1)/d_Q] d_Q$.

Proof [9]: Let us introduce the following differential operators in δ_y and δ_z: $D_y = y P(\delta_y)$, $D_z = z^{-1} Q(\delta_z)$, so that $[D_y, D_z] = 0$. For $\zeta = yz$, we have

$$c_n \zeta^n = D_y^{-1} D_z c_{n+1} \zeta^{n+1} \tag{5.4}$$

We formally put $c_{-1} = c_{-2} = \cdots = 0$, so that the formula (5.4) is true for all integers n [note that $Q(x)$ has a zero at $x = 0$]. We now put $A(x) = \sum_{i=0}^{n} x^{n-i} g_{n-i}(1/c_{i+r-n})$, with the definition of g_{n-i} given in the statement of Theorem 5.2. We denote $A(x) = \sum_{i=0}^{n} x^{n-i} a_i$. Then we have

$$A(x) f(\zeta) = \left(\sum_{i=0}^{n} x^{n-i} a_i \right) \left(\sum_{\ell=0}^{\infty} c_\ell \zeta^\ell \right) = \sum_{s=0}^{\infty} S_s(x, \zeta) \tag{5.5}$$

where $S_s(x, \zeta)$ are polynomials homogeneous in x, ζ of degree s:

$$S_s(x, \zeta) = \sum_{i=0}^{n} x^{n-i} a_i \zeta^{i+s-n} c_{i+s-n}$$

Then, according to the definition of a_i, $a_i = g_{n-i}(1/c_{i+r-n})$. Hence for $s = r$ we have

$$S_r(x, \zeta) = \sum_{i=0}^{n} x^{n-i} g_{n-i} \zeta^{i+r-n}$$

$$= \sum_{i=0}^{n} x^{n-i} (-1)^i \binom{n}{i} P(i + b) \cdots P(i + r - n - 1) (yz)^{i+r-n}$$

We note that by the definition of the operator $D_y = y P(\delta_y)$, $D_y y^m = P(m) y^{m+1}$. Hence $S_r(x, \zeta)$ can immediately be identified with $D_y^{r-n-b} (x - yz)^n y^b z^{r-n}$. Indeed,

$$S_r(x, yz) = D_y^{r-n-b} (x - yz)^n y^b z^{r-n}$$

Hence, according to (5.4),

$$S_{r-j}(x, yz) = D_y^{-j} D_z^j S_r(x, yz): \quad j = 0, \ldots, r$$

Algebra for Diophantine Equations

In particular,

$$S_{r-j}(x,yz) = D_y^{r-n-b-j} D_z^j (x - yz)^n y^b z^{r-n}$$

For $j = 0, 1, \ldots, r - n - b$, $D_y^{r-n-b-j} D_z^j$ is a differential operator of the order $\deg(P)(r - n - b - j) + \deg(Q)j$. Hence $S_{r-j}(x,\zeta)$ has a zero at $x = \zeta$ of an order at least $n - [\deg(P)(r - n - b - j) + \deg(Q)j] = n - \{[(n - 1)/d_Q]\deg(P) + j[\deg(Q) - \deg(P)]\} \geq n - [(n - 1)/d_Q]d_Q$ for $j \leq r - n - b = [(n - 1)/d_Q]$, $d_Q = \max\{\deg(P),\deg(Q)\}$. Hence we have

$$\left(\frac{d}{d\zeta}\right)^k S_{r-j}(x,\zeta)\Big|_{x=\zeta} = 0 \tag{5.6}$$

for all $j = 0, 1, \ldots, r - n - b = [(n - 1)/d_Q]$ and $k = 0, 1, \ldots, (n - 1) - [(n - 1)/d_Q]d_Q$. From (5.5) and (5.6) it follows that for every $k = 0, 1, \ldots, (n - 1) - [(n - 1)/d_Q]d_Q$,

$$A(x)\delta_\zeta^k f(\zeta)\Big|_{\zeta=x} = \sum_{s=0}^{\infty} \delta_\zeta^k S_s(x,\zeta)\Big|_{\zeta=x}$$

$$= \sum_{s=0}^{n+b-1} \delta_\zeta^k S_s(x,\zeta)\Big|_{\zeta=x} + \sum_{s=r+1}^{\infty} \delta_\zeta^k S_s(x,\zeta)\Big|_{\zeta=x} \tag{5.7}$$

We remark now that $\delta_\zeta^k S_s(x,\zeta)\Big|_{\zeta=x} = E_s^{(k)} x^k$, so Theorem 5.2 follows from (5.7).

Now let $n = d_Q N$. Then $r = N(1 + d_Q) + b - 1$, and $(n - 1) - [(n - 1)/d_Q]d_Q = d_Q - 1$. Then for $k = 0, 1, \ldots, d_Q - 1$, we have

$$A(x)\delta_x^k f(x) - B_k(x) = O(x^{N(1+d_Q)+b})$$

where $\deg(A) = n = d_Q N$, $\deg(B_k) = d_Q N + b - 1$.

COROLLARY 5.4 Let $m = \max\{q + 1, p\}$, and $f_1(x) = 1$, $f_2(x) = $ $m - 1$ is the order of the differential equation (5.1) satisfied by $f(x)$. Then, in the notations of Theorem 5.2 for $n = (m - 1)N$, the polynomials $A(x)$, $B_0(x)$, $\ldots$, $B_{m-2}(x)$ are Padé approximants of type II to the functions $f_1(x)$, $\ldots$, $f_m(x)$ with weights

$$(N + b - 1, N, \ldots, N)$$

We can easily make a transfer from $\delta_x^k f(x)$: $k = 0, \ldots, m - 2$ to the usual contiguous functions. In the standard hypergeometric notations let

$$f(x) = {}_pF_q\left(x \middle| \begin{matrix} a_1, \ldots, a_p \\ b_1, \ldots, b_q \end{matrix}\right)$$

and

$$f(x,k) = {}_pF_q\left(x \middle| \begin{matrix} a_1 + k, a_2, \ldots, a_p \\ b_1, \ldots, b_q \end{matrix}\right)$$

for $k \leq 0$. Then

$$f(x,k) = [a \cdots (a + k - 1)]^{-1}(\delta_x + a) \cdots (\delta_x + a + k - 1)f(x)$$

for $k \geq 1$. Hence the system of explicit Padé approximants of type II in the diagonal case for the system of functions

$$1, {}_pF_q\left(x \middle| \begin{matrix} a_1, \ldots, a_p \\ b_1, \ldots, b_q \end{matrix}\right), {}_pF_q\left(x \middle| \begin{matrix} a_1 + 1, a_2, \ldots, a_p \\ b_1, \ldots, b_q \end{matrix}\right), \ldots,$$

$$ {}_pF_q\left(x \middle| \begin{matrix} a_1 + q, a_2, \ldots, a_p \\ b_1, \ldots, b_q \end{matrix}\right)$$

is given by a linear combination of polynomials $B_0(x), \ldots, B_q(x)$, and by $A(x)$:

COROLLARY 5.4   Let $m = \max\{q + 1, p\}$, and $f_1(x) = 1$, $f_2(x) = {}_pF_q\left(x \middle| \begin{matrix} a_1, \ldots, a_p \\ b_1, \ldots, b_q \end{matrix}\right)$, $f_i(x) = {}_pF_q\left(x \middle| \begin{matrix} a_1 + i - 2, a_2, \ldots, a_p \\ b_1, \ldots, b_q \end{matrix}\right)$: $i = 2, \ldots, m + 1$.

Then the system of diagonal Padé approximants to $f_1(x), \ldots, f_m(x)$ corresponding to weights $(N, \ldots, N)$ is given by the polynomials $A(x)$ and $B_0(x)$, $\ldots, B_{m-1}(x)$ from Theorem 5.2 for $n = Nm$, for example, $A_1(x|N, \ldots, N) = A(x)$.

We also present the system of Padé approximants of type II to functions

$$1 \text{ and } \delta_x^k f(\alpha_i x): k = 0, 1, \ldots, d_Q - 1; \, i = 1, \ldots, d$$

where $\alpha_1, \ldots, \alpha_d$ are distinct nonzero complex numbers and d_Q is the order of the differential equation (5.1) satisfied by the generalized hypergeometric function $f(x)$. Now let

$$\prod_{i=1}^{d} (\alpha_i - t) = \sum_{j=0}^{d} \sigma_j t^j$$

and

Algebra for Diophantine Equations

$$\left(\sum_{j=0}^{d} \sigma_j t^j\right)^{d_Q k} = \sum_{\ell=0}^{k d d_Q} p_{\ell,k} t^\ell$$

where $p_{\ell,k}$ are symmetric functions in $\alpha_1, \ldots, \alpha_d$.

THEOREM 5.5 Let $d_Q = \max\{\deg(P), \deg(Q)\}$ be the order of equation (5.1) satisfied by the generalized hypergeometric function $f(x) = \sum_{n=0}^{\infty} c_n x^n$. For given N, we define

$$A(x) = \sum_{i=0}^{Ndd_Q} x^{Ndd_Q - i} p_{i,N} P(i+1) \cdots P(i+N-1) \frac{1}{c_{i+N}}$$

Let us consider the sequence of numbers $E_{s,j}^{(k)}$: $s = 0, 1, 2, \ldots$; $j = 1, \ldots, d$; $k = 0, 1, \ldots, d_Q - 1$:

$$E_{s,j}^{(k)} = \sum_{i=0}^{Ndd_Q} (i + s - Ndd_Q)^k p_{i,N} P(i+1) \cdots P(i+N-1) \frac{c_{i+b-Ndd_Q}}{c_{i+N}}$$

(where we put $c_{-1} = c_{-2} = \cdots = 0$). Then for

$$B_{k,j}(x) = \sum_{s=0}^{Ndd_Q} E_{s,j}^{(k)} x^s \alpha_j^s$$

$k = 0, 1, \ldots, d_Q - 1$; $j = 1, \ldots, d$, we have

$$A(x) \delta_x^k f(\alpha_j x) - \alpha_j^k B_{k,j}(x) = x^{N(1+dd_Q)+1} \sum_{s=N(1+dd_Q)+1}^{\infty} x^{s-N(1+dd_Q)-1}$$
$$\times \alpha_j^k E_{s,j}^{(k)}$$

Hence $A(x)$ and $B_{k,j}(x)$ are Padé approximants of type II to $dd_Q + 1$ functions: 1 and $\delta_x^k f(\alpha_j x)$: $k = 0, 1, \ldots, d_Q - 1$; $j = 1, \ldots, d$, for equal weights $(N, N, \ldots, N)$. Alternatively, we obtain Padé approximations of type II to functions

$$1 \text{ and } {}_pF_q\left(\alpha_j x \bigg| \begin{matrix} a_1 + i, a_2, \ldots, a_p \\ b_1, \ldots, b_q \end{matrix}\right)$$

for $j = 1, \ldots, d$ and $i = 0, 1, \ldots, \max\{p, q+1\} - 1$.

Proof: We follow the method of Theorem 5.2. Namely, we have for $\zeta = yz$, $A(x) f(\zeta) = \sum_{s=0}^{\infty} S_s(x, \zeta)$, and for $r = N + Ndd_Q$,

$$S_r(x,\zeta) = D_y^{N-1}\left[\prod_{i=1}^{d} (\alpha_i x - yz)^{d_Q N} y z^N\right]$$

Then the order of zero of $S_{r-\lambda}(x,\zeta)$ at $\zeta = \alpha_j x$ is at least $d_Q N - (r - Ndd_Q - 1)d_Q = d_Q$ for all $j = 1, \ldots, d$ provided that $r - Ndd_Q - 1 \geq \lambda \geq 0$, or $N - 1 \geq \lambda \geq 0$. This proves Theorem 5.5.

Theorems 5.2 and 5.5 are immediately generalized to basic generalized hypergeometric functions:

$$_p\varphi_r\left(z \left|\begin{array}{c}\alpha_1,\ldots,\alpha_p \\ \beta_1,\ldots,\beta_r\end{array}\right.\right) = \sum_{n=0}^{\infty} \frac{(\alpha_1)_{q,n} \cdots (\alpha_p)_{q,n}}{(\beta_1)_{q,n} \cdots (\beta_r)_{q,n} (a)_{q,n}} z^n$$

for $(a)_{q,n} = (1-a)(1-aq)(1-aq^2)\cdots(1-aq^{n-1})$, $(a)_{q,0} = 1$.

This way we get diagonal Padé approximants of type II to

$$_p\varphi_r\left(zd_j \left|\begin{array}{c}\alpha_1 q^i,\ldots,\alpha_p \\ \beta_1,\ldots,\beta_r\end{array}\right.\right): j = 1, \ldots, d; i = 0, 1, \ldots, \max\{p, r+1\} - 1$$

REMARK 5.6 Theorem 5.2 provides an unusually simple expression for Padé approximants (of type II) for the (generalized) hypergeometric series $f(x) = \Sigma_{n=0}^{\infty} c_n x^n$ in terms of c_n. Small change in notations $f(x) = \Sigma_{n=0}^{\infty} (c_n/n!^d) x^n (d = d_Q)$, $r = (d+1)N$, $b = 1$, represents the Padé approximant to $f(x)$ [and to $\delta_x f(x), \ldots, \delta_x^{d-1} f(x)$] as

$$A_N(x) = x^N \sum_{i=0}^{dN} x^{-i} \binom{dN}{i} \binom{N+i}{i}^d \frac{1}{c_i}$$

This form of Padé approximants seems to be totally independent of properties of coefficients c_i and runs contrary to well-known expressions for coefficients of Padé approximants in terms of Henkel determinants of c_i (complex rational expressions of c_i related to characters of S_N—cf. [41-42]). Theorem 5.2 shows that a simple expression given above for $A_N(x)$ in terms of c_i holds for all N only when $f(x)$ is a generalized hypergeometric function; that is, the ratios of the consecutive coefficients in the expansion of $f(x)$ is a rational function of index of degree at most d. To understand why Padé approximants to $f(x)$ have good arithmetic properties, it is enough to notice that whenever $f(x) = \Sigma_{n=0}^{\infty} c_n x^n$ is a generalized hypergeometric function, the function $\Sigma_{n=0}^{\infty} (1/c_n) x^n$ is also a generalized hypergeometric function.

6. MEASURES OF DIOPHANTINE APPROXIMATIONS FOR VALUES OF SPECIAL G-FUNCTIONS

As identities of Section 4 show, such constants of classical analysis as π can be represented as rapidly convergent series with integral or nearly integral coefficients evaluated at a (rational) point very close to the origin. In fact, many constants of classical analysis and geometry are represented in the same way, including periods of algebraic varieties, considered as values of solutions to Picard-Fuchs deformation equations. An appropriate class of functions, called G-functions, had been introduced by Siegel [1] as a companion to the class of E-functions that describe entire functions such as Bessel functions. While for E-functions Siegel proved strong transcendence and algebraic independence results [1,3], for G-functions, which are the only interesting functions in geometric applications, Siegel outlined the only program that had been left unfulfilled until very recently. We present these new results from Refs. 12 and 14 fulfilling the program of Ref. 1 together with more specific results for special G-functions--generalized hypergeometric ones.

We use the standard notations of algebraic number theory. For an algebraic number α and a complete set $\{\alpha_1 = \alpha, \ldots, \alpha_d\}$ of numbers algebraically conjugate to α, we denote by $|\bar{\alpha}| = \max\{|\alpha_1|, \ldots, |\alpha_d|\}$ the size of α. Also, we denote by $\text{den}\{\alpha_0, \ldots, \alpha_n\}$ the common denominator of $\alpha_0, \ldots, \alpha_n$.

DEFINITION 6.1 (Siegel) Let $f(x) = \sum_{n=0}^{\infty} a_n x^n$ be a solution of a linear differential equation over $\bar{\mathbb{Q}}(x)$. $f(x)$ is called a G-function if $a_n \in \bar{\mathbb{Q}}$ and there exists a constant $C > 0$ such that $\overline{|a_n|} \leq C^n$ and the common denominator of $a_0, \ldots, a_n$ is at most C^n.

REMARK 6.2 In fact, all coefficients a_n of the expansion of $f(x)$ belong to a fixed algebraic number field K. The field K is generated by the coefficients of a linear differential equation from $\bar{\mathbb{Q}}(x)[d/dx]$, satisfied by $f(x)$ and by the first few a_n. For the purposes of this chapter we assume that $K = \mathbb{Q}$. Such a reduction to the $K = \mathbb{Q}$ case can be achieved by considering simultaneously with $f(x) = \sum_{n=0}^{\infty} a_n x^n \in K[[x]]$ all functions $f^{(\sigma)}(x) = \sum_{n=0}^{\infty} a_n^{(\sigma)} x^n$ with $\sigma(a_n) = a_n^{(\sigma)}$ for isomorphic embeddings $\sigma: K \to \mathbb{C}$. Then the functions $\text{Sym}(f^{(\sigma)}(x): \sigma: K \to \mathbb{C})$ for all symmetric combinations Sym of $f^{(\sigma)}(x)$ are already G-functions with $K = \mathbb{Q}$ (i.e., one can replace $\bar{\mathbb{Q}}$ by $\mathbb{Q}$ in Definition 6.1).

Obviously, algebraic functions are G-functions, because by Eisenstein's theorem, for an algebraic function $f(x)$ with the expansion $f(x) = \sum_{n=0}^{\infty} a_n x^n$ with $a_n \in \bar{\mathbb{Q}}$, the common denominator of $a_0, \ldots, a_n$ divides AB^n for appropriate integers A, B. Also, the class of G-functions is closed under integration, addition, multiplication, and differentiation. In particular, solutions of Picard-Fuchs equations, including hypergeometric ${}_{m+1}F_m\left(\begin{matrix}a_1,\ldots,a_{m+1}\\b_1,\ldots,b_m\end{matrix}\bigg| x\right)$-functions with rational a_i, b_j also belong to the class of G-functions.

In Ref. 1 Siegel proposed considering those (rational) points x close to the origin such that for $x = p/q$, $|x| < |q|^{-\varepsilon}$ or even $|x| < \exp(-\log|q|^{1/2+\varepsilon})$ for $\varepsilon > 0$ and large q, $|q| \geq q_0(\varepsilon)$. The first positive results in the direction suggested by Siegel immediately revealed the need of additional geometric assumptions on G-functions or rather on linear differential equations these functions satisfied. These additional conditions (Refs. 35-37) are closely connected with the global nilpotent conditions and the Grothendieck conjecture (see Refs. 14, 38, and 39 for a review of p-adic properties of p-adic differential equation). To present these conditions we look at a system $f_1(x), \ldots, f_n(x)$ of G-functions satisfying a system of the first-order linear differential equations over $\bar{\mathbb{Q}}(x)$:

$$\frac{d}{dx} \underline{f}^t = A \underline{f}^t \tag{6.1}$$

$\underline{f} = (f_1(x), \ldots, f_n(x))$, and $A = A(x) \in M(n, \bar{\mathbb{Q}}(x))$. Typically, $f_i(x) = (d/dx)^{i-1} f(x)$, where $f(x)$ satisfies a scalar linear differential equation over $\bar{\mathbb{Q}}(x)$ of order n.

Let $D(x)$ be a (polynomial) denominator of the elements of the matrix $A = A(x)$. We can iterate (6.1) k times and get

$$\frac{d}{dx}^k \underline{f}^t = A_k(x) \underline{f}^t \tag{6.2}$$

$\mod(d/dx - A(x))$. A strong additional condition on G-functions $f_1(x), \ldots, f_n(x)$ that is called a (G,C)-function condition is the demand that there exists a constant $C_2 \geq 1$ such that in (6.2) for all $N \geq 1$, the common denominator of all coefficients of polynomial entries of matrices $(1/k!)A_k(x)D(x)^k$: $k = 0, 1, \ldots, N$ is bounded by C_2^N. The (G,C)-function condition is a strong geometric constraint, because it implies, in particular, global nilpotence of the system (6.2). The global nilpotence of (6.2) (see Refs. 37-39) means that the p-curvature operator

$$\psi_p = \left(\frac{d}{dx} I - A\right)^p \mod p$$

is nilpotent for almost all prime p. The global nilpotence condition is a very restrictive one, and one can even speculate that all globally nilpotent equations can be solved in quadratures (multiple integrals of algebraic functions).

In Refs. 12 and 14 we used Padé approximations of the second kind (see the definitions in Section 5) to prove the (G,C)-function conditions and global nilpotence of linear differential equations (6.2) satisfied by G-functions. These systems of Padé-type approximations of the second kind take the following form:

$$R_i(x) \stackrel{def}{=} Q(x) f_i(x) - P_i(x): i = 1, \ldots, n$$

where $P_1(x), \ldots, P_n(x), Q(x)$ are polynomials in x of degrees at most D, and such that

$$\mathrm{ord}_{x=0} R_i(x) \geq D + \frac{D}{n} - \varepsilon D: i = 1, \ldots, n$$

One of our main results in this area is the following:

THEOREM 6.3 [12,14] Let $f_1(x), \ldots, f_n(x)$ be a system of G-functions satisfying a system of first-order linear differential equations (6.1) over $\bar{\mathbb{Q}}(x)$. If $f_1(x), \ldots, f_n(x)$ are linearly independent over $\bar{\mathbb{Q}}(x)$, the functions $f_1(x), \ldots, f_n(x)$ are (G,C)-functions.

This result and Padé-type approximations of the second kind applied directly to solutions of (6.1) imply general results of linear independence sought by Siegel:

THEOREM 6.4 [14] Let $f_1(x), \ldots, f_n(x)$ be G-functions with rational coefficients in their Taylor expansions, satisfying a first-order linear differential system (6.1) over $\mathbb{Q}(x)$, and such that functions $1, f_1(x), \ldots, f_n(x)$ are linearly independent over $\mathbb{Q}(x)$. Then for any $\varepsilon > 0$ and arbitrary rational $r = a/b$ with (rational) integers a and b such that $|b|^\varepsilon \geq c_3 |a|^{(n+1)(n+\varepsilon)}$, $r \neq 0$, the numbers $1, f_1(r), \ldots, f_n(r)$ are linearly independent over $\mathbb{Q}$; and for arbitrary rational integers $H_0, H_1, \ldots, H_n$ we have

$$\left| H_0 + H_1 f_1\left(\frac{a}{b}\right) + \cdots + H_n f_n\left(\frac{a}{b}\right) \right| > H^{-n-\varepsilon}$$

with $H = \max\{|H_0|,\ldots,|H_n|\}$, when $H \geq h_0$. Here $c_3 = c_3(f_1,\ldots,f_n,\varepsilon) > 0$, $h_0 = h_0(f_1,\ldots,f_n,\varepsilon,r) > 0$ are effective constants.

In general, we have

$$\left| H_0 + H_1 f_1\left(\frac{a}{b}\right) + \cdots + H_n f_n\left(\frac{a}{b}\right) \right| > H^{\lambda-\varepsilon}$$

with $\lambda = -n \log|b|/\log|b/a^{n+1}|$, whenever $|b| \geq c_4|a|^{n+1}$ and $H \geq h_1$ in the notations above, with effective constants $c_4 = c_4(f_1,\ldots,f_n,n) > 0$ and $h_1 = h_1(f_1,\ldots,f_n,n,r) > 0$.

Similar results hold for values of algebraic functions. For proofs, see Refs. 12 and 14.

THEOREM 6.5 Let $f_1(x), \ldots, f_n(x)$ be G-functions satisfying matrix first-order linear differential equations (6.1) over $\bar{\mathbb{Q}}(x)$, and such that functions $1, f_1(x), \ldots, f_n(x)$ are algebraically independent over $\bar{\mathbb{Q}}(x)$. Then for any $t \geq 1$ there exists an effective constant $c_5 = c_5(f_1,\ldots,f_n,t) > 0$ such that for any algebraic number $\xi \neq 0$ if degree $\leq t$, it follows from

$$|\xi| < \exp[-c_5(\log|\bar{\xi}|)^{4n/(4n+1)}] \qquad (6.3)$$

that numbers

$$1, f_1(\xi), \ldots, f_n(\xi)$$

are not related by an algebraic relation of degree $\leq t$ over $\mathbb{Q}(\xi)$. (Similar results are valid for values of the G-function in several archimedean and nonarchimedean metrics.)

Let us denote, for an arbitrary G-function $f(x) = \sum_{n=0}^{\infty} a_n x^n$ ($a_n \in K$: $n = 0, 1, \ldots$) and a place v of K, by $f^{(v)}(x)$ the function defined on the completion of K corresponding to v. For example, for the one (ith) archimedean place v_i corresponding to the embedding $\alpha \to \alpha^{(i)}$ of $K \to \mathbb{C}$, $f^{(v_i)}(x) = \sum_{n=0}^{\infty} a_n^{(i)} x^n$. For a nonarchimedean place v, $f^{(v)}(x)$ is defined on the completion K_v of K. For example, the value of $f^{(v)}(x)$ at $x \in K_v$ is a v-adic number from K_v and it can be different from the value of $f(x)$, even when $x \in K \subset K_v$. Because $f(x)$ is a G-function, every function $f^{(v)}(x)$ has a nonzero radius of convergence in K_v.

In these notations, the results of Theorems 6.4 and 6.5 hold for any function $f^{(v)}(x)$. Namely, we have

Algebra for Diophantine Equations

REMARK If in Theorem 6.5 we consider K containing $\mathbb{Q}(\xi)$ and the field of coefficients of expansions of $f_i(x)$: $i = 1, \ldots, n$, then the results of Theorem 6.5 hold for functions $f_1^{(v)}(x), \ldots, f_n^{(v)}(x)$ instead of $f_1(x), \ldots, f_n(x)$. One has only to replace $|\xi|$ in (6.3) by $|\xi|_v$ and (complex) numbers $f_1(\xi), \ldots, f_n(\xi)$ by v-adic ones: $f_1^{(v)}(\xi), \ldots, f_n^{(v)}(\xi)$ from K_v. The constant c_3 then depends on v as well. Similarly, in Theorem 6.4, under the assumptions $|a/b|_v \leq c_3 \max(|a|,|b|)^{\varepsilon/(n+1)(n+\varepsilon)-1}$, the v-adic numbers 1, $f_1^{(v)}(r), \ldots, f_n^{(v)}(r)$ are linearly independent over $\mathbb{Q}$ and $|H_0 + H_1 f_1(r) + \cdots + H_n f_n(r)|_v > H^{-n-\varepsilon}$ with $H = \max(|H_0|,\ldots,|H_n|) \geq h_0$ and c_3, h_0 depending on v.

In many applications one needs strong results of the form of Theorem 6.4 when functions $f_1(x), \ldots, f_n(x)$ satisfy linear differential equations of an arbitrary order over $\bar{\mathbb{Q}}(x)$, without increasing the exponent in the lower bounds of linear forms in values of G-functions. For this we use methods of graded Padé approximations proposed by us in Refs. 2, 10, and 11. One of our results is the following:

THEOREM 6.6 Let $f_1(x), \ldots, f_n(x)$ be G-functions with rational number coefficients of Taylor expansions, satisfying linear differential equations of arbitrary order, over $\mathbb{Q}(x)$. Then for any $\varepsilon > 0$ and a rational number $r = a/b$, with integers a and b such that $|b| \geq c_7 |a|^{n(n-1+\varepsilon)}$, $c_7 = c_7(f_1, \ldots, f_n, \varepsilon) > 0$, we have the following lower bound for linear forms in $f_1(r), \ldots, f_n(r)$. For arbitrary nonzero rational integers $H_1, \ldots, H_n$ and $H = \max\{|H_1|,\ldots,|H_n|\}$, if $H_1 f_1(r) + \cdots + H_n f_n(r) \neq 0$, then

$$|H_1 f_1(r) + \cdots + H_n f_n(r)| > |H_1 \cdots H_n|^{-1} H^{1-\varepsilon}$$

provided that $H \geq c_8$ with $c_8 = c_8(f_1,\ldots,f_n,r,\varepsilon) > 0$, and effective $c_7 > 0$, $c_8 > 0$.

Under the same assumptions on r, for functions 1, $f_1(x), \ldots, f_n(x)$ linearly independent over $\mathbb{Q}(x)$ and arbitrary rational integers $q, q_1, \ldots, q_n$, we have

$$|q_1 \cdots q_n|^{1+\varepsilon} \|q_1 f_1(r) + \cdots + q_n f_n(r)\| > 1$$

and

$$|q|^{1+\varepsilon} \|qf_1(r)\| \cdots \|qf_n(r)\| > 1$$

provided that $|q_1 \cdots q_n| > c_9$ and $|q| > c_9$. Here $\|\cdot\|$ is the distance to the nearest integer, and $c_9 = c_9(f_1,\ldots,f_n,r,\varepsilon) > 0$ is an effective constant.

In all results above we can also explicitly exhibit the dependence of constants c_8 and c_9 on r, namely on $|b|$. This is of particular importance in our applications to algebraic functions, where $r = 1/b$ with varying b. For example, under the assumptions of Theorem 6.6 we have for rational integers $H_1, \ldots, H_n$:

$$|H_1 f_1(r) + \cdots + H_n f_n(r)| > |b|^{-n} H^{\lambda - \varepsilon}$$

with $\lambda = -(n-1)\log|b|/\log|b/a^n|$, $H = \max\{|H_0|,\ldots,|H_n|\}$ provided that $H \geq c_{10}(f_1,\ldots,f_n,\varepsilon)$.

The proof of Theorem 6.6 is based on graded Padé approximation methods developed by the authors in Refs. 10 and 11. The essence of these methods consists in simultaneous approximations of all elements of graded submodules in Picard-Vessiot extensions of $\mathbb{C}(x)$ generated by linear differential equations satisfied by $f_1(x), \ldots, f_n(x)$. Namely, under the assumptions of Theorem 6.6, let functions $f_i(x)$ satisfy scalar linear differential equations over $\mathbb{Q}(x)$ of orders k_i: $i = 1, \ldots, n$. We introduce auxiliary variables $c_{i,j}$ ($j = 1, \ldots, k_i$; $i = 1, \ldots, n$) and $\bar{c}_i = (c_{i,1},\ldots,c_{i,n})$: $i = 1, \ldots, n$.

DEFINITION 6.7 (Graded Padé Approximations) Let $P_i(x|\bar{c})$: $i = 1, \ldots, n$ be polynomials in x of degrees at most D and in $\bar{c} = (\bar{c}_1,\ldots,\bar{c}_n)$, homogeneous in each group of variables $\bar{c}_j = (c_{j,1},\ldots,c_{j,k_j})$: $j \neq i$ of degree N, and in variables $\bar{c}_i = (c_{i,1},\ldots,c_{i,k_i})$ of degree $N - 1$: $i = 1, \ldots, n$. Let the remainder function

$$R(x|\bar{c}) = \sum_{i=1}^n P_i(x|\bar{c}) \left[\sum_{j=1}^{k_i} c_{i,j} f_i^{(j-1)}(x) \right]$$

have a zero at $x = 0$ of order at least t for any choice of $\bar{c} = (\bar{c}_1,\ldots,\bar{c}_n)$. If $t \geq nD - \varepsilon D$ and

$$t \geq \sum_{i=1}^n D \frac{\binom{N + k_i - 2}{k_i - 1}}{\binom{N + k_i - 1}{k_i - 1}} - \varepsilon \frac{D}{\prod_{i=1}^n \binom{N + k_i - 1}{k_i - 1}}$$

then $P_i(x|\bar{c})$ are called Padé approximants and $R(x|\bar{c})$ is called a remainder function in the ε-graded Padé approximation problem with weight D of level N.

Algebra for Diophantine Equations

The remainder functions in the graded Padé approximations are then specialized for values of G-functions in the proof of Theorem 6.6. This is very similar to proofs of our earlier results on exponential functions (Refs. 2 and 10) and general E-functions [11].

Results like Theorem 6.6 have interesting applications to Ramanujan-like identities of Section 4. There we showed that π can be represented up to rational or quadratic factors as a value of a combination of hypergeometric (thus G-) functions at points like $12^3/j(\tau)$ whenever $\tau \in \mathbb{Q}(\sqrt{-d})$ for one-class discriminants $-d$. Combining identities like (4.7) and (4.9) with Theorem 6.6 we arrive at a remarkable

MOCK PROPOSITION If there are infinitely many one-class (negative quadratic) discriminants, then *all* elements of the field $\mathbb{Q}(\pi^2)$ transcendental over $\mathbb{Q}$ have measures of irrationality with exponents $-2 - \varepsilon$ for all $\varepsilon > 0$.

Unfortunately, this "Roth" theorem for all polynomials in π^2 is not proved because degrees of numbers $j(\tau)$ grow to infinity as $d \to \infty$. Nevertheless, we can obtain some good bounds on the measures of irrationality of π and its factors from identities of Section 4. For this, however, we apply explicit Pade approximations of the second kind from Section 5 instead of general graded Padé approximation techniques. We present a few such examples where new dense systems of rational approximations demanded by Lemma 2.5 are constructed by methods of Section 5.

EXAMPLE 6.8 (Sequence P_n/Q_n approximating $\pi\sqrt{2}$) The first expression for the denominators Q_n is the following:

$$Q_n = \sum_{i=0}^{3n} \binom{3n}{i}\binom{n+i-1}{i}^3 \frac{i!^4}{(ri)!} (-1)^i \cdot 396^{4i}$$

Numerators P_n and denominators Q_n satisfy five-term linear recurrence. An improvement over this dense system of approximations to $\pi\sqrt{2}$ in the style of Section 2 leads to the following measure of irrationality for $\pi\sqrt{2}$:

$$|q\pi\sqrt{2} - p| > |q|^{-15.67}$$

for a sufficiently large $|q|$.

These rational approximations to $\pi\sqrt{2}$ are closely related to Ramanujan's famous rapidly convergent representation of $1/(2\pi\sqrt{2})$:

$$\frac{1}{2\pi\sqrt{2}} = \frac{1103}{99^2} + \frac{27,493}{99^6} \cdot \frac{1}{2} \cdot \frac{1 \cdot 3}{4^2} + \frac{53,883}{99^{10}} \cdot \frac{1 \cdot 3}{2 \cdot 4} \cdot \frac{1 \cdot 3 \cdot 5 \cdot 7}{4^2 \cdot 8^2} + \cdots$$

(in customary Ramanujan notation [25]).

EXAMPLE 6.9   Another sequence of approximations given by generalized hypergeometric polynomials is P_n/Q_n, where denominators Q_n are the following:

$$Q_n = \sum_{i=0}^{3n} \binom{3n}{i}\binom{n+i-1}{i}^3 \frac{(3i)!i!^3}{(6i)!} (640,320)^{3i}$$

Example 6.9 follows from the identity (4.10) and Padé approximations of Section 5. Improving Q_n in the style of Section 2 gives a new sequence P_n/Q_n.

The irrational number, which the sequence P_n/Q_n approximates well, is

$$\alpha = \frac{\sqrt{640,320}}{\pi}$$

The measure of irrationality of α which follows from this sequence of P_n/Q_n turns out to be

$$|q\alpha - p| > |q|^{-11.1\ldots}$$

The SCRATCHPAD system was also used in this example to find other closed expressions for Padé approximants of the second kind to particular hypergeometric functions related to the elliptic integrals of the first and second kind.

EXAMPLE 6.10   (Number π^2)   In a search of rational approximations P_n/Q_n to π^2 we start with Apéry's sequence of appriximations, where the denominators Q_n have the form

$$Q_n = \sum_{i=0}^{n} \binom{n}{i}^2 \binom{n+i}{i}$$

and denominators of rational numbers P_n divide $\text{lcm}\{1,\ldots,n\}^2$.

We used SCRATCHPAD here to test which sums of binomials can serve as denominators in the rational approximation problem for best approximations. The first natural candidate for better rational approximations P_n/Q_n to π^2 leads to the following choice of the denominators Q_n:

$$Q_n = \sum_{j=0}^{N_2} \sum_{\substack{i=0 \\ i \geq N_2 - 2j}}^{N_1} (-1)^j \binom{N_1}{i}\binom{N_1}{i+2j-N_2}\binom{N_2}{j}\binom{N_0+i+j}{n}$$

for parameters N_0, N_1, N_2.

Algebra for Diophantine Equations

Optimal choice of parameters N_0, N_1, and N_2 giving the best approximations P_n/Q_n to π^2 with rational P_n, whose denominators divide $\text{lcm}\{1,\ldots,n\}^2$ is the following: $N_0 = [0.9n]$, $N_1 = [0.95n]$, and $N_2 = [0.1n]$.

The measure of irrationality of π^2 it implies is the following:

$$|q\pi^2 - p| > |q|^{-9.31} \qquad |q| \geq q_6$$

We have checked various expressions of denominators Q_n in rational approximations P_n/Q_n to π^2. One of the general forms of expressions of Q_n is the following:

$$Q_n = \sum_{m=0}^{N_2+N_3} 3 \sum_{i=0}^{m} \sum_{j=0}^{m} \binom{N_3}{i}\binom{N_3}{j}\binom{N_2}{m-i}\binom{N_2}{m-j} 2^{i+j} \sum_{k=0}^{N_1} 1 \binom{N_0+m}{n-k}\binom{N_1}{k} 2^k$$

for appropriate parameters N_0, N_1, N_2, N_3. (The case $N_0 = N_2 = n$, $N_1 = N_3 = 0$ gives us Apéry's Q_n.)

In this expression for Q_n, one optimal choice of parameters N_0, N_1, N_2, N_3 from the point of view of constants of approximations of π^2 is the following:

$$N_0 = [0.78n] \qquad N_1 = [0.12n] \qquad N_2 = [0.96n] \qquad N_3 = [0.14n]$$

This choice of parameters gives rise as in Section 2 to dense approximations to π^2 that imply the following measure of irrationality:

$$|q\pi^2 - p| > |q|^{-6.901\ldots}$$

for rational integers p and q with $|q| > q_7$.

More complicated expressions for denominators Q_n, represented as a sum of products of *eight* binomial coefficients, give a still slightly better measure of irrationality of π^2:

$$|q\pi^2 - p| > |q|^{-6.5}$$

whenever $|q| \geq q_8$. Hence, for π we get

$$|q\pi - p| > |q|^{-14}$$

for $|q| \geq q_9$.

Note that sometimes exceptionally good approximations do exist. For π^4 we have $|22\pi^4 - 2143| < 22^{-4.89\ldots}$

7. THE ARITHMETIC AND ANALYTIC ASYMPTOTICS OF PADÉ APPROXIMANTS

The four classes of $_3F_2$ hypergeometric functions (which are squares of $_2F_1$ hypergeometric functions) that arise in the Ramanujan quadratic period relations of Section 4 are the following:

$$_3F_2\left(\begin{matrix} 1/2, 1/6, 5/6 \\ 1, 1 \end{matrix} \bigg| x\right) = \sum_{n=0}^{\infty} \frac{(6n)!}{(3n)! n!^3} \left(\frac{x}{12^3}\right)^n \tag{7.1}$$

$$_3F_2\left(\begin{matrix} 1/4, 3/4, 1/2 \\ 1, 1 \end{matrix} \bigg| x\right) = \sum_{n=0}^{\infty} \frac{(4n)!}{n!^4} \left(\frac{x}{4^4}\right)^n \tag{7.2}$$

$$_3F_2\left(\begin{matrix} 1/2, 1/2, 1/2 \\ 1, 1 \end{matrix} \bigg| x\right) = \sum_{n=0}^{\infty} \frac{(2n)!^3}{n!^6} \left(\frac{x}{2^6}\right)^n \tag{7.3}$$

$$_3F_2\left(\begin{matrix} 1/3, 2/3, 1/2 \\ 1, 1 \end{matrix} \bigg| x\right) = \sum_{n=0}^{\infty} \frac{(3n)!}{n!^3} \frac{(2n)!}{n!^2} \left(\frac{x}{3^3 \cdot 2^2}\right)^n \tag{7.4}$$

Let us apply the general construction of Section 5 of Padé approximants of type II to these four classes of $_3F_2$ functions that arise in Ramanujan-type representations of algebraic multiples of $1/\pi$. We start with one of the most interesting examples: $_3F_2\left(\begin{matrix} 1/4, 1/2, 3/4 \\ 1, 1 \end{matrix} \bigg| x\right)$. This function arises naturally as a square of the period of an elliptic curve, for example,

$$\left[\frac{2K(k_N)}{\pi}\right]^2 = (1 + k_N^2)^{-1} \cdot {}_3F_2\left(\begin{matrix} 1/4, 1/2, 3/4 \\ 1, 1 \end{matrix} \bigg| 4(g_N^{12} + g_N^{-12})^{-2}\right)$$

for $k_N^2 = k^2(\sqrt{-N})$, $g_N = 2^{-1/4} f_1(\sqrt{-N})$. The corresponding (generalized) hypergeometric function is (7.2)

$$F_2(x) \stackrel{\text{def}}{=} {}_3F_2\left(\begin{matrix} 1/4, 1/2, 3/4 \\ 1, 1 \end{matrix} \bigg| x\right) = \sum_{n=0}^{\infty} \frac{(4n)!}{n!^4} \left(\frac{x}{4^4}\right)^n$$

The system of Padé approximants of type II from Section 5 provides us with the simultaneous approximations to $F_2(x)$ and $\delta_x F_2(x) = x F_2'(x)$, necessary to obtain measures of diophantine approximations to multiples of $1/\pi$. In notations of Theorem 5.2 and Corollary 5.3, we choose $d_Q = 3$, $Q(x) = x^3$, $P(x) = (x + 1/4)(x + 1/2)(x + 3/4)$, $m = 4$, $b = 1$, $n = 3N$, $r = 4N$ for a parameter N. With the notation $z = x/4^4$, $F_2(z) = \sum_{n=0}^{\infty} [(4n)!/n!^4] z^n$, both functions $F_2(z)$ and $\delta_z F_2(z) = z F_2'(z)$ [as well as $\delta_z^2 F(z)$] are Padé approximated at $z = 0$ by Padé approximants of type II with the common denominator closely related to the following polynomial:

Algebra for Diophantine Equations

$$A_N^{(2)}(z) = \sum_{i=0}^{3N} \binom{3N}{i}\binom{N+i}{i}^3 \frac{i!^4}{(4i)!}(-z)^{3N-i} \tag{7.5}$$

One of the main problems in applications of Padé approximation methods is the determination of uses and denominators of the coefficients of Padé approximants. We use the techniques proposed in Ref. 40 and determine the common denominator of coefficients in (7.5).

We have for the coefficients of $A_N^{(2)}(z)$ in (7.5),

$$\binom{3N}{i}\binom{N+i}{i}^3 \frac{i!^4}{(4i)!} = \frac{(3N-i+1)\cdots(3N)}{i!} \frac{[(N+1)\cdots(N+i)]^3}{i!^3}$$
$$\times \frac{i!^4}{(4i)!} = \frac{[(3N-i+1)\cdots(3N)][(N+1)\cdots(N+i)]^3}{(4i)!} \tag{7.6}$$

for $0 \le i \le 3N$.

The denominator $\Delta_N = \Delta_N^{(2)}$ of coefficients of $A_N^{(2)}(x)$ can consist of primes $p \le 12N$. Of those, the contribution of primes $p \le O(\sqrt{N})$ to Δ_N is at most $e^{O(N)}$ (see the arguments in Ref. 40). Thus we look at $p \ge O(\sqrt{N})$.

We look now at primes p lying in the intervals

$$\left[\frac{12}{j+1}N, \frac{12}{j}N\right] \quad \text{for } j = 0, 1, \ldots$$

Thus for a nonnegative integer j we look at primes p such that

$$jp \le 12N < (j+1)p \tag{7.7}$$

Let us look now at the integer k_s such that

$$k_s p \le sN < (k_s + 1)p$$

Here we easily get $k_{12} = j$, $k_6 = [j/2]$, $k_4 = [j/3]$, $k_3 = [j/4]$, $k_2 = [j/6]$, $k_1 = [j/12]$; that is, we put

$$k_s = \left[\frac{j}{t}\right] \quad \text{if } st = 12 \tag{7.8}$$

To determine the $\nu_p(\cdot)$ of numbers (7.6) we look at three new variables: t_1, t_3, and k such that

$$k_1 p < N \le (k_1 + 1)p \le \cdots \le (k_1 + t_1)p \le N + i < (k_1 + t_1 + 1)p \tag{7.9}$$

and

$$(k_3 - t_3)p < 3N - i + 1 \le (k_3 - t_3 + 1)p \le \cdots \le k_3 p \le 3N < (k_3 + 1)p \tag{7.10}$$

and

$$kp \leq 4i < (k+1)p \tag{7.11}$$

Here, according to our assumption on p, $p \geq O(\sqrt{N})$, we get from (7.9)-(7.11), respectively:

$$\nu_p([(N+1)\cdots(N+i)]) = t_1 \tag{7.12}$$

$$\nu_p([(3N-i+1)\cdots(3N)]) = t_3 \tag{7.13}$$

$$\nu_p((4i)!) = k \tag{7.14}$$

Consequently, $-\nu_p(\Delta_N) = \min\{3t_1 + t_3 - k : 0 \leq i \leq 3N\}$, where $\min\{\cdot : \cdot\}$ is taken over all t_1, t_3, k in (7.9)-(7.11) for i in $0, \ldots, 3N$.

Let us write all inequalities on i and in (7.9)-(7.11):

$$0 \leq i \leq 3N \tag{7.15a}$$

$$(k_1 + t_1)p - N \leq i < (k_1 + t_1 + 1)p - N \tag{7.15b}$$

$$3N - (k_3 - t_3 + 1)p + 1 \leq i < 3N - (k_3 - t_3)p + 1 \tag{7.15c}$$

$$\frac{kp}{4} \leq i < \frac{k+1}{4}p \tag{7.15d}$$

In addition, there are obvious conditions

$$t_1 \geq 0 \quad t_3 \geq 0 \quad k \geq 0 \tag{7.16}$$

Let us eliminate i from (7.15) by looking at the consistency conditions:

$$k_1 + t_1 \leq k_4 \tag{7.17}$$

$$t_3 \leq k_3 \tag{7.18}$$

$$k \leq k_{12} \tag{7.19}$$

Further consistency conditions are:

$$k_1 + k_3 + t_1 - t_3 \leq k_4 : \quad (b) + (c) \tag{7.20}$$

$$k_4 + 1 \leq k_1 + k_3 + t_1 - t_3 + 1 + 1 : \quad (b) + (c) \tag{7.21}$$

$$4(k_1 + t_1) - (k+1) \leq k_4 : \quad (b) + (d) \tag{7.22}$$

$$k_4 + 1 \leq 4(k_1 + t_1 + 1) - k : \quad (b) + (d) \tag{7.23}$$

$$k_{12} + 1 \leq 4(k_3 - t_3 + 1) + (k+1) : \quad (c) + (d) \tag{7.24}$$

$$4(k_3 - t_3) + k \leq k_{12} : \quad (c) + (d) \tag{7.25}$$

Thus from (7.20) and (7.21) we obtain

$$k_4 - k_1 - k_3 - 1 \leq t_1 - t_3 \leq k_4 - k_1 - k_3 \tag{7.26}$$

Algebra for Diophantine Equations 67

From (7.22) and (7.23) we derive

$$4k_1 - k_4 - 1 + 4t_1 \le k \le 4k_1 - k_4 + 3 + 4t_1 \tag{7.27}$$

and from (7.24) and (7.25),

$$k_{12} - 4k_3 + 4t_3 - 4 \le k \le k_{12} - 4k_3 + 4t_3 \tag{7.28}$$

One checks directly the consistency of conditions (7.16)-(7.19) and (7.26)-(7.28), which are considered as inequalities on k for given parameters t_1 and t_3 (obvious, but a tedious comparison). Because of the consistency, and because $-\nu_p(\Delta_N) = \min\{-k + 3t_1 + t_3\}$, one can take k as large as inequalities (7.19), (7.27), and (7.28) allow to maximize $\nu_p(\Delta_N)$. This means that we can put

$$k = \min\{k_{12}, 4k_1 - k_4 + 3 + 4t_1, k_{12} - 4k_3 + 4t_3\} \tag{7.29}$$

or symbolically, $k = \min\{j; A, B\}$ according to the order of (7.29).

Which of A or B is larger depends on which side of the inequality (7.26) $t_1 - t_2$ is equal to.

REMARK If $t_1 - t_3 = k_4 - k_1 - k_3$, then $A > B$. If $t_1 - t_3 = k_4 - k_1 - k_3 - 1$, then $B > A$.

Proof: $A - B = 4k_1 - k_{12} + 4k_3 - k_4 + 3 + 4(t_1 - t_3) = 4k_1 - k_{12} + 4k_3 - k_4 + 3 + 4\left[k_4 - k_1 - k_3 + \begin{pmatrix}0\\1\end{pmatrix}\right] = 3k_4 - k_{12} + 3 + 4\begin{pmatrix}0\\1\end{pmatrix} = 3[j/3] - j + \begin{pmatrix}3\\-1\end{pmatrix}$.

In the first case, $A - B > 0$; in the second, $A - B < 0$.

Let us consider separately the two possibilities in the inequality (7.26).

First Case: Let $t_1 - t_3 = k_4 - k_1 - k_3 - 1$. Then, according to (ac)-(ad), $t_1 = k_4 - k_1 - P$, $t_3 = k_3 + 1 - P$ for $P \ge 0$. Also according to (7.16), $P \le \min\{k_3 + 1, k_4 - k_1\}$. According to the Remark, $A < B$. In (7.29) we get $A = 4k_1 - k_4 + 3 + 4k_4 - 4k_1 - 4P = 3[j/3] + 3 - 4P$, and $j = k_{12}$. We have $j \ge A$ ($= 3[j/3] + 3 - 4P$) only when $P \ge 1$. Thus if $P = 0$, we have in (7.29), $k = j$, and if $P \ge 1$, $k = 3k_4 + 3 - 4P$. Thus if $P = 0$, we have

$$-\nu_p(\Delta_N) \le -j + 3k_4 - 3k_1 + k_3 + 1 \tag{7.30}$$

If $P \ge 1$, we have

$$-\nu_p(\Delta_N) \le k_3 - 3k_1 - 2 \tag{7.31}$$

Clearly, among numbers on the right sides of (7.30) and (7.31), the number on the right side of (7.31) is smaller [i.e., the corresponding value of $\nu_p(\Delta_N)$ is larger]. Consequently, whenever $P \geq 1$ is possible (i.e., $k_3 + 1 \geq 1$, $k_4 - k_1 \geq 1$) we choose $-\nu_p(\Delta_N)$ according to (7.31). Thus for $j \geq 3$,

$$-\nu_p(\Delta_N) \leq k_3 - 3k_1 - 2 \tag{7.32}$$

and for $j = 1, 2$,

$$p\nu_p(\Delta_N) \leq -j + 1 \tag{7.33}$$

Second Case: Let $t_1 - t_3 = k_4 - k_1 - k_3$. Then, according to (ac)-(ad), $t_1 = k_4 - k_1 - P$, $t_3 = k_3 - P$ for $P \geq 0$ [if consistent with (7.16)]. According to the Remark, $B < A$. We have in (7.29), $B = k_{12} - 4k_3 + 4t_3 = j - 4P$, $k_{12} = j$. Thus $B \leq j$, and in (7.29), $k = B = j - 4P$. Consequently, in this case,

$$-\nu_p(\Delta_N) = -j + 3k_4 - 3k_1 + k_3 \tag{7.34}$$

Again, the number on the right side of (7.32) is smaller than the one in (7.34) for $j \geq 3$, and the number on the right in (7.34) is smaller than that in (7.33) for $j = 1, 2$. Consequently, we have

$$-\nu_p(\Delta_N) = \begin{cases} k_3 - 3k_1 - 2 & \text{for } j \geq 3 \\ -j & \text{otherwise} \end{cases} \tag{7.35}$$

Our description of the common denominator $\Delta_N^{(2)} = \Delta_N$ of the coefficients of (7.5) can be summarized as follows:

COROLLARY 7.1 For $N \geq 0$, a common denominator $\Delta_N^{(2)}$ of numbers (7.6), a prime $p \geq O(\sqrt{N})$, lying in the interval $jp \leq 12N < (j+1)p$, we have

$$\nu_p(\Delta_N^{(2)}) = 2 + 3\left[\frac{j}{12}\right] - \left[\frac{j}{4}\right]$$

when $j \geq 3$, or $\nu_p(\Delta_N^{(2)}) = j$ otherwise. Consequently, the asymptotics of $\Delta_N^{(2)}$ as $N \to \infty$ is the following:

$$\frac{1}{N} \log \Delta_N^{(2)} \to 12 + 2\left[\psi\left(\frac{4}{3}\right) - \psi(1)\right] + \psi\left(\frac{2}{3}\right) - \psi\left(\frac{1}{3}\right)$$
$$= 18 - 2 \ln 6 - \ln 3$$

The first part of Corollary 7.1 follows directly from (7.35). The second part is a consequence of the first part and the definition of $\psi(z)$—the logarithmic derivative of the Γ-function (see the arguments in Ref. 40).

Algebra for Diophantine Equations

Let us look now at the next $_3F_2$ function in (7.3). In this case we are looking at the square of the Legendre expression of a period of an elliptic curve:

$$\left[\frac{2K(k)}{\pi}\right]^2 = {}_3F_2\left(\begin{array}{c}1/2, \ 1/2, \ 1/2 \\ 1, \ 1\end{array}\bigg| 4k^2(1-k^2)\right)$$

Let us look at a system of Padé approximants of type II that Padé approximate the functions $F_3(z) = \sum_{n=0}^{\infty} [(2n)!^3/n!^6] z^n$ (i.e., $z = x/2^6$) and in which $\delta_z F_3(z) = z F_3'(z)$ [as well as $\delta_z^2 F_3(z)$] are Padé approximated at $z = 0$ by Padé approximants closely related to the polynomial

$$A_N^{(3)}(z) = \sum_{i=0}^{3N} \binom{3N}{i}\binom{N+i}{i}^3 \frac{i!^6}{(2i)!^3} (-z)^{3N-i} \qquad (7.36)$$

We have for the coefficients of $A_N^{(3)}(z)$ in (7.41),

$$\binom{3N}{i}\binom{N+i}{i}^3 \frac{i!^6}{(2i)!^3} = \frac{(3N-i+1)\cdots(3N)}{i!} \frac{[(N+1)\cdots(N+i)]^3}{i!^3} \frac{i!^6}{(2i)!^3}$$

$$= \frac{[(3N-i+1)\cdots(3N)][(N+1)\cdots(N+i)]^3 (i!)^2}{(2i)!^3}$$

(7.37)

for $0 \leq i \leq 3N$.

The denominator $\Delta_N^{(3)} = \Delta_N$ of coefficients $A_N^{(3)}(z)$ can consist of primes $p \leq 6N$. Of those, the contribution of primes $p \leq O(\sqrt{N})$ is at most $e^{O(N)}$ (see the arguments in Ref. 40). Thus we look at $p \geq O(\sqrt{N})$.

We look now at primes p lying in the intervals

$$\left[\frac{6}{j+1} N, \ \frac{6}{j} N\right) \quad \text{for } j = 0, 1, \ldots$$

Thus for a nonnegative integer j we look at primes p such that

$$jp \leq 6N < (j+1)p \qquad (7.38)$$

Let us look now at the integer k_s such that

$$k_s p \leq sN < (k_s + 1)p$$

Here we easily get $k_6 = j$, $k_3 = [j/2]$, $k_2 = [j/3]$, $k_1 = [j/6]$, that is, we put

$$k_s = \left[\frac{j}{t}\right] \quad \text{if } st = 6 \qquad (7.39)$$

To determine the $\nu_p(\cdot)$ of numbers (7.37) we look at the new variables t_1, t_3, ℓ, and k such that

$$k_1 p < N \leq (k_1 + 1)p \leq \cdots \leq (k_1 + t_1)p \leq N + i < (k_1 + t_1 + 1)p \tag{7.40}$$

$$(k_3 - t_3)p < 3N - i + 1 \leq (k_3 - t_3 + 1)p \leq \cdots$$
$$\leq k_3 p \leq 3N < (k_3 + 1)p \tag{7.41}$$

and

$$\ell p \leq i < (\ell + 1)p \tag{7.42}$$

$$kp \leq 2i < (k + 1)p \tag{7.43}$$

Here, according to our assumption on p, $p \geq O(\sqrt{N})$, we get from (7.40) and (7.41), respectively:

$$\nu_p([(N + 1) \cdots (N + i)]) = t_1 \tag{7.44}$$

$$\nu_p([(3N - i + 1) \cdots (3N)]) = t_3 \tag{7.45}$$

$$\nu_p(i!) = \ell \qquad \nu_p((2i)!) = k \tag{7.46}$$

Consequently, $-\nu_p(\Delta_N) = \min\{3t_1 + t_3 + 2\ell - 3k : 0 \leq i \leq 3N\}$ where $\min\{\cdot:\cdot\}$ is taken over all t_1, t_3, k for i in 0, ..., 3N.

Let us write all inequalities on i and in (7.40)-(7.43):

$$0 \leq i \leq 3N \tag{7.47a}$$

$$(k_1 + t_1)p - N \leq i < (k_1 + t_1 + 1)p - N \tag{7.47b}$$

$$3N - (k_3 - t_3 + 1)p + 1 \leq i < 3N - (k_3 - t_3)p + 1 \tag{7.47c}$$

$$\ell p \leq i < (\ell + 1)p \tag{7.47d}$$

$$\frac{kp}{2} \leq i < \frac{k+1}{2} p \tag{7.47e}$$

In addition, there are obvious conditions

$$t_1 \geq 0 \qquad t_3 \geq 0 \qquad k \geq 0 \qquad \ell \leq 0 \tag{7.48}$$

We eliminate i, looking at these inequalities as consistency conditions. The resulting inequalities on t_1, t_3, ℓ, and k (see the proof of Corollary 7.1) can be represented more easily by denoting $j = 12r + s$ for $0 \leq s < 11$. Then we obtain the following inequalities on k and ℓ in terms of t_1 and t_3:

$$-\left[\frac{s}{6}\right] - 1 + 2t_1 \leq k \leq -\left[\frac{s}{6}\right] + 1 + 2t_1 \tag{7.49}$$

Algebra for Diophantine Equations

$$\left[\frac{s}{2}\right] - 2\left[\frac{s}{4}\right] + 2t_3 - 2 \leq k \leq \left[\frac{s}{2}\right] - 2\left[\frac{s}{4}\right] + 2t_3 \tag{7.50}$$

$$t_1 - 1 \leq \ell \leq t_1 \tag{7.51}$$

$$2\ell \leq k \leq 2\ell + 1 \tag{7.52}$$

The bounds for t_1 and t_3 are the following:

$$\begin{aligned} t_1 &\leq \tfrac{1}{2}\left(6r + \left[\tfrac{s}{2}\right] - 1\right) \\ t_1 &\leq 3r + \left[\tfrac{s}{3}\right] - \left[\tfrac{s}{12}\right] \\ t_3 &\leq 3r + \left[\tfrac{s}{4}\right] \end{aligned} \tag{7.53}$$

We look at the minimal value of $-\nu_p(\Delta_N) = 3t_1 + t_3 + 2\ell - 3k$ subject to conditions (7.49)-(7.53) for given $s = 0, \ldots, 11$. Simple analysis shows that $-\nu_p(\Delta_N) = -2$ for $s = 0, 1, 4, 5$; $-\nu_p(\Delta_N) = -3$ for $s = 2, 3$ and $\nu_p(\Delta_N) = 0$ for $s = 6, \ldots, 11$. Thus we arrive at

COROLLARY 7.2 For $N \geq 0$, and for a common denominator $\Delta_N^{(3)}$ of coefficients of the polynomial $A_N^{(3)}(z)$ of (7.36),

$$\sum_{p \geq 0 (\sqrt{N})} \nu_p(\Delta_N^{(3)}) = N\left[2 \sum_{r \geq 1}\left(\frac{1}{r} - \frac{1}{r + 1/6}\right) + 3 \sum_{r \geq 1}\left(\frac{1}{r + 1/6} - \frac{1}{r + 2/6}\right)\right.$$
$$\left. + 2 \sum_{r \geq 1}\left(\frac{1}{r + 2/6} - \frac{1}{r + 3/6}\right)\right] \tag{7.54}$$

Consequently, the asymptotics of $\Delta_N^{(3)}$ as $N \to \infty$ is as follows:

$$\frac{1}{N} \log \Delta_N^{(3)} \to 2\left[\psi\left(\tfrac{7}{6}\right) - \psi(1)\right] + 3\left[\psi\left(\tfrac{2}{6}\right) - \psi\left(\tfrac{1}{6}\right)\right] + 2\left[\psi\left(\tfrac{3}{6}\right) - \psi\left(\tfrac{2}{6}\right)\right]$$

$$= 12 - \ln 4 + \frac{\pi}{2}\left(\sqrt{3} - \frac{1}{\sqrt{3}}\right)$$

$$= 12.42750592 \ldots$$

Let us finally turn to the $_3F_2$ function in (7.1). This function corresponds to $x = 12^3/J$, J being the value of the modular invariant at an appropriate singular modulus. We look at the Padé approximants of type II corresponding to the hypergeometric function in (7.1): $F_1(z) = \sum_{n=0}^{\infty} [(6n)!/(3n)!n!^3] z^n$ (with $z = x/12^3$) and $\delta_z F_1(z)$ [and $\delta_z^2 F_1(z)$]. The expression for the corresponding Padé approximant (the denominator part) was constructed in Section 5 for $r = 3N$, $b = 1$, $d_Q = 3$. Again the related polynomial is:

$$A_N^{(1)}(z) = \sum_{i=0}^{3N} \binom{3N}{i} \binom{N+i}{i} \frac{i!^3 (3i)!}{(6i)!} (-z)^{3N-i} \qquad (7.55)$$

(with $z = x/12^3$ as above).

As in the proofs of Corollaries 7.1 and 7.2, we determine the common denominator Δ_N of the coefficient in (7.5). Again the main contribution to Δ_N comes from primes $p \geq O(\sqrt{N})$.

We look at coefficients of (7.55):

$$\frac{[(3N-i+1)\cdots(3N)][(N+1)\cdots(N+i)]^3 (3i)!}{i!\,(6i)!}: \quad 0 \leq i \leq 3N \qquad (7.56)$$

$$k_s p \leq sN < (k_s + 1)p \qquad (7.57)$$

and (for a given i: $0 \leq i \leq 3N$)

$$\ell_\alpha p \leq \alpha i < (\ell_\alpha + 1)p \qquad (7.58)$$

As above,

$$k_1 p < N \leq (k_1 + 1)p \leq \cdots \leq (k_1 + t_1)p \leq N + i < (k_1 + t_1 + 1)p \qquad (7.59)$$

$$(k_3 - t_3)p < 3N - i + 1 \leq (k_3 - t_3 + 1)p \leq \cdots$$

$$\leq k_3 p \leq 3N < (k_3 + 1)p \qquad (7.60)$$

Thus, for a given p, $O(\sqrt{N}) \leq p < 18N$:

$$\nu_p([(N+1)\cdots(N+i)]) = t_1$$

$$\nu_p([(3N-i+1)\cdots(3N)]) = t_3$$

$$\nu_p(i!) = \ell_1$$

$$\nu_p((3i)!) = \ell_3$$

$$\nu_p((6i)!) = \ell_6$$

Thus, $-\nu_p(\Delta_N^{(3)}) = \min\{3t_1 + t_3 + \ell_3 - \ell_1 - \ell_6:$ for i in $0, \ldots, 3N\}$.

The inequalities on i are the following:

$$0 \leq i \leq 3N \qquad (7.61a)$$

$$(k_1 + t_1)p - N \leq i < (k_1 + t_1 + 1)p - N \qquad (7.61b)$$

$$3N - (k_3 - t_3 + 1)p + 1 \leq i < 3N - (k_3 - t_3)p + 1 \qquad (7.61c)$$

$$\ell_\alpha p \leq \alpha i < (\ell_\alpha + 1)p \qquad (7.61d)$$

Algebra for Diophantine Equations

In (7.61d) it is enough to take $\alpha = 6$. The consistency conditions on i in (7.61) are

$$t_1 \leq k_4 - k_1$$
$$t_3 \leq k_3 \tag{7.62}$$
$$\ell_6 \leq k_{18}$$

More consistency conditions:

$$k_4 - k_1 - k_3 - 1 \leq t_1 - t_3 \leq k_4 - k_1 - k_3 \tag{7.63}$$
$$6k_1 - k_6 - 1 + 6t_1 \leq \ell_6 \leq 6k_1 - k_6 + 5 + 6t_1 \tag{7.64}$$
$$k_{18} - 6k_3 - 6 + 6t_3 \leq \ell_6 \leq k_{18} - 6k_3 + 6t_3 \tag{7.65}$$

To determine the minimal values of $\nu_p(\Delta_N^{(1)})$ we look at $j = 36r + s$, when $k_1 = \ell$, $k_3 = 3\ell + [s/12]$, $k_4 = 4\ell + [s/9]$, $k_6 = 6\ell + [s/6]$, $k_{18} = 18\ell + [s/2]$. Then the inequalities on t_1, t_3, and ℓ_6 [in addition to (7.62)] are

$$\left[\frac{s}{9}\right] - \left[\frac{s}{12}\right] - 1 \leq t_1 - t_3 \leq \left[\frac{s}{9}\right] - \left[\frac{s}{12}\right] \tag{7.66}$$

$$-\left[\frac{s}{6}\right] - 1 + 6t_1 \leq \ell_6 \leq -\left[\frac{s}{6}\right] + 5 + 6t_1 \tag{7.67}$$

$$\left[\frac{s}{2}\right] - 6\left[\frac{s}{12}\right] - 6 + 6t_3 \leq \ell_6 \leq \left[\frac{s}{2}\right] - 6\left[\frac{s}{12}\right] + 6t_3 \tag{7.68}$$

for $-\nu_p(\Delta_N^{(1)}) = \min\{3t_1 + t_3 + [\ell_6/2] - [\ell_6/6] - \ell_6\}$. Looking at $s = 0$, ..., 35 we arrive at the following determination of $\nu_p(\Delta_N^{(1)})$ according to $j \mod 36$ (in fact, $j \mod 18$):

COROLLARY 7.3 For $N \geq 0$ and the common denominator $\Delta_n^{(1)}$ of coefficients of Padé approximants $A_N^{(1)}(z)$ in (7.55) we have the following determination of $\nu_p(\Delta_N^{(1)})$ for primes $p \geq 0(\sqrt{N})$:

$$\nu_p(\Delta_N^{(1)}) = 1 \quad \text{for } \frac{18N}{3} < p \leq 18N$$

$$\nu_p(\Delta_N^{(1)}) = 2 \quad \text{for } \frac{18N}{6} < p \leq \frac{18N}{3}$$

$$\nu_p(\Delta_N^{(1)}) = 2 \quad \text{for } \frac{18N}{18r + 6} < p \leq \frac{18N}{18r}$$

$$\nu_p(\Delta_N^{(1)}) = 1 \quad \text{for } \frac{18N}{18r + 12} \leq p \leq \frac{18N}{18r + 6}$$

$$\nu_p(\Delta_N^{(1)}) = 1 \quad \text{for } \frac{18N}{18r + 15} < p \leq \frac{18N}{18r + 13}$$

and $\nu_p(\Delta_N^{(1)}) = 0$ otherwise. Consequently, we have the following asymptotics of $\Delta_N^{(1)}$ as $N \to \infty$:

$$\frac{1}{N} \log \Delta_n^{(1)} \to 18 + 2\left|\psi\left(\frac{24}{18}\right) - \psi(1)\right| + \psi\left(\frac{12}{18}\right) - \psi\left(\frac{6}{8}\right) + \psi\left(\frac{15}{18}\right) - \psi\left(\frac{13}{18}\right)$$

$$= 20.9720213848\ldots$$

Estimates of the denominator of the coefficients of the polynomial

$$A_N^{(4)}(z) = \sum_{i=0}^{3N} \binom{3N}{i}\binom{N+i}{i}^3 \frac{i!^5}{(2i)!\,(3i)!} (-z)^{3N-i}$$

similar to the hypergeometric function in (7.4) can also be proved.

Note that the denominator of the Padé approximant is simultaneously the denominator of coefficients of all other Padé approximants $B_{N,j}(z)$: because the corresponding (normalized) hypergeometric series have integral coefficients.

To apply Padé approximations from Section 5, in addition to the integrality statements of the form of Corollaries 7.1-7.3, we also need precise results on the convergence of these Padé approximations. Namely, to prove the bound on the measure of diophantine approximations in the framework of Lemma 2.5, we need asymptotical results from Ref. 9 (see Theorems 4.1 and 5.1 in Ref. 9). To be more specific, we consider Padé approximations of the second kind to the following system of functions: 1 and $\delta_x^j f(x)$ for $j = 0, \ldots, q$, where

$$f(x) = {}_{q+1}F_q\left(x \left| \begin{matrix} a_1, \ldots, a_{q+1} \\ b_1, \ldots, b_q \end{matrix}\right.\right) \tag{7.69}$$

We look at Padé approximants with weights $(N,\ldots,N)$. This corresponds to the choice of $n = (q+1)N$, $b = 1$ in Theorem 5.2 and Corollary 5.3. From Theorem 5.2 it follows that the Padé approximant $A(x)$ is a generalized hypergeometric polynomial of degree $n = (q+1)N$ of the form

$$A(x) = x^{(q+1)N} {}_{q+2}F_{q+1}\left(x^{-1} \left| \begin{matrix} -(q+1)N, N+1, N+b_1, \ldots, N+b_q \\ a_1+1, \ldots, a_{q+1}+1 \end{matrix}\right.\right)$$

The other Padé approximants $B_0(x), \ldots, B_q(x)$ for which $A(x)(\delta_x^j f(x)) - B_j(x) = O(x^{(q+1)N+1})$, $j = 0, \ldots, q$, are also generalized hypergeometric polynomials of degree N. They can be defined in various ways, but the best definition for arithmetic applications is the following one:

Algebra for Diophantine Equations 75

$$B_j(x) = [A(x)\delta_x^j f(x)]_N: \quad j = 0, \ldots, q$$

Here $[u(x)]_N$ denotes a Taylor polynomial of degree N of a power series expansion $u(x)$ at $x = 0$ (i.e., sum of terms of $u(x)$ up to Nth order). These Padé approximations are convergent within the radius of convergence of $f(x)$ (i.e., for $|x| < 1$). In fact, they converge outside this disk as well, but we do not need this in most of our applications. The leading term in the asymptotic expansion of $A(x) = A_N(x)$ and $R_{N,j}(x) \stackrel{def}{=} A_N(x)\delta_x^j f(x) - B_j(x)$: $j = 0, \ldots, q$ can be determined from the generalized hypergeometric integral representations whenever $x \neq 0$ is fixed and is within a unit circle and $N \to \infty$. This leading term is determined by a solution of a trinomial algebraic equation of degree $q + 2$. Combining Theorems 4.1 and 5.1 of Ref. 9, we obtain

PROPOSITION 7.4 For fixed (real) x, $0 < |x| < 1$ and $N \to \infty$, the asymptotics of the Padé approximants $A_N(x)$ and $B_{N,j}(x)$, and the asymptotics of the remainder function $R_{N,j}(x) = A_N(x)\delta_x^j f(x) - B_{N,j}(x)$, $j = 0, \ldots, q$, in a Padé approximation problem of type II to 1, $\delta_x^j f(x): j = 0, \ldots, q$ for $f(x)$ from (7.69) with weights $(N, \ldots, N)$ are determined by the roots of a trinomial

$$(q + 1)t^{q+2} - (q + 2)t^{q+1} + x = 0$$

If t_m is the largest absolute value root of this equation, then

$$|A_N(x)|^{1/N} \sim ((q + 1)t_m^{q+2})^{(q+1)}$$

as $N \to \infty$. [Here the degree of $A_N(x)$ is $(q + 1)N$.] If t^m is the smallest absolute value root of this trinomial, then

$$|R_{N,j}(x)|^{1/N} \sim ((q + 1)t^{m(q+2)})^{(q+1)}$$

for $j = 0, \ldots, q$. For $x \to 0$, we have the following dependencies of t_m, t^m on x: $t_m = [(q + 2)/(q + 1)] + O(x)$, $t^m = (q + 2)^{-1/(q+1)} x^{1/(q+1)} [1 + O(x^{1/(q+1)})]$, which is consistent with asymptotics of Padé approximants and remainder functions. The asymptotics of $|B_{N,j}(x)|^{1/N}$ are the same as those of $|A_N(x)|^{1/N}$ for $N \to \infty$ and $0 < |x| < 1$.

8. DIFFERENTIAL ALGEBRA COMPUTATIONS

One of the advantages of different computer algebra systems relates to their broad possibilities for differential algebra calculations. Such a capability is particularly important for studies of nonlinear completely integrable systems defined by partial differential equations. Such tasks as a search for algebraic conservation laws of bounded degrees are easily accessible to symbolic algebra programs. A simple problem arising from the completely integrable three-dimensional systems is related to the "universal form" of completely integrable lattice equations. One such equation is Hirota's discrete analog of a generalized Toda equation on a function $f = f(\bar{z})$ [59]:

$$(\lambda_2 - \lambda_3)f(\bar{z} + \bar{\lambda}_1)f(\bar{z} + \bar{\lambda}_2 + \bar{\lambda}_3) + (\lambda_3 - \lambda_1)f(\bar{z} + \bar{\lambda}_2)f(\bar{z} + \bar{\lambda}_1 + \bar{\lambda}_3)$$
$$+ (\lambda_1 - \lambda_2)f(\bar{z} + \bar{\lambda}_3)f(\bar{z} + \bar{\lambda}_1 + \bar{\lambda}_2) = 0 \quad (8.1)$$

In this equation one can interpret $\bar{z}$ as a three-dimensional vector and $\bar{\lambda}_1$, $\bar{\lambda}_2$, $\bar{\lambda}_3$ as three directions corresponding to constants λ_1, λ_2, λ_3. One of the interesting results shown by Hirota is that all known two- and three-dimensional completely integrable systems can be recovered in various limits (when translations are approximating partial derivatives) from equation (8.1). This result is correct for equations of the so-called isospectral deformation type, which includes Korteweg-deVries, nonlinear Schrödinger, sine-Gordon equations, and all their higher analogs. Computer algebra systems are very convenient for getting various limits of (8.1) and finding nontrivial equations among them. Even the first nontrivial equation is nontrivial indeed. It is equivalent to the well-known Kadomtzev-Petviashvili (KP) equation, taking the familiar form

$$12 u_{yy} = \frac{\partial}{\partial x}(12 u_t - 12 u u_x - u_{xxx}) \quad (8.2)$$

after the transformation [in (8.1)]

$$u = \frac{\partial^2}{\partial x^2} \log f$$

There is one more reason why study of the equation (8.1) and its limits is mathematically important. Equation (8.1) is a form of a law of addition for the Riemann θ-function $f(\bar{z})$ coming from algebraic curves. This form of the law of addition was originally studied by Fay and Gunning. Recently, it was rigorously proved that equation (8.1) and its limit form (8.2) provide a compact answer to the Schottky problem on determining jacobians among all abelian varieties. Computer algebra systems seem to be an ideal instrument for the explicit solution of Schottky problems (at least those involving not too large genuses).

The interest in completely integrable systems that developed among mathematicians and physicists during the last dozen years has led to an ultimate question: how we can, looking at an algebraic ordinary or partial differential equation, recognize its complete integrability character and find its first integrals or conservation laws. A complete answer to this question presumably should be in the form of a computer algebra program to determine the question of complete integrability once an equation is introduced into a computer.

There are several sufficient conditions for a system of ordinary or partial differential equations to represent completely integrable (hamiltonian) systems. Except for trivial systems, most of the criteria point to nonlinear equations solvable via that or another version of the inverse scattering method. This leads, of course, to another problem: When is the equation solvable via the inverse scattering method? One can again try computer algebra to answer this new question.

However, for a system to be completely integrable, a different condition appears to be necessary. This is the so-called "Painlevé test," first proposed by Painlevé for ordinary differential equations. Painlevé proposed to describe all ordinary differential equations of bounded order (for Painlevé himself it was the second order) having no movable singularities in their solutions. This led to the famous Painlevé list of six irreducible equations of the second order [50]. Studies made by Painlevé and his students at the turn of the century attracted attention about a decade ago in connection with self-similarity solutions of completely integrable partial differential equations. Recently, Painlevé tests were developed and applied successfully to a variety of multidimensional nonlinear partial differential equations as a means of efficient determination of their complete integrability properties. Such tests (for a particular example, see Ref. 60) are the beneficiaries of the power of computer algebra systems.

ACKNOWLEDGMENT

This work was supported in part by the U.S. Air Force.

REFERENCES

1. C. L. Siegel, Über einige Anwendungen diophantischer Approximationen, Abh. Preuss., Akad. Wiss. Phys. Math. Kl., *1* (1929).

2. G. V. Chudnovsky, *Contributions to the Theory of Transcendental Numbers* (Mathematical Surveys and Monographs 19), American Mathematical Society, Providence, R.I., 1984.

3. A. Baker, *Transcendental Number Theory*, Cambridge University Press, Cambridge, 1979.

4. D. V. Chudnovsky and G. V. Chudnovsky, *Multidimensional Hermite Interpolation and Padé Approximation* (Lecture Notes in Mathematics 925), Springer-Verlag, Berlin, 1982, pp. 271-298.

5. G. V. Chudnovsky, *Rational and Padé Approximations to Solutions of Linear Differential Equations and the Monodromy Theory* (Lecture Notes in Physics 126), Springer-Verlag, Berlin, 1980, pp. 136-169.

6. G. V. Chudnovsky, Padé approximation and the Riemann monodromy problem, in *Bifurcation Phenomena in Mathematical Physics and Related Topics*, D. Reidel, Boston, 1980, pp. 448-510.

7. G. V. Chudnovsky, *Hermite-Padé Approximations to Exponential Functions and Elementary Estimates of the Measure of Irrationality of* π (Lecture Notes in Mathematics 925), Springer-Verlag, Berlin, 1982, pp. 299-322.

8. G. V. Chudnovsky and D. V. Chudnovsky, Multisoliton formula for completely integrable two-dimensional systems, Lett. Nuovo Cimento *25* (1979), 263-265.

9. G. V. Chudnovsky, Padé approximations to the generalized hypergeometric functions I, J. Math. Pures Appl. (Paris) *58* (1979), 445-476.

10. D. V. Chudnovsky and G. V. Chudnovsky, *Padé Approximations to Solutions of Linear Differential Equations and Applications to Diophantine Analysis* (Lecture Notes in Mathematics 1052), Springer-Verlag, Berlin, 1980, pp. 85-167.

11. G. V. Chudnovsky, On some applications of diophantine approximations, Proc. Natl. Acad. Sci. U.S.A. *81* (1984), 1926-1930.

12. G. V. Chudnovsky, On applications of diophantine approximations, Proc. Natl. Acad. Sci. U.S.A. *81* (1984), 7261-7265.

13. W. M. Schmidt, *Diophantine Approximations* (Lecture Notes in Mathematics 785), Springer-Verlag, Berlin, 1980.

14. D. V. Chudnovsky and G. V. Chudnovsky, *Applications of Padé Approximations to Diophantine Inequalities in Values of G-Functions* (Lecture Notes in Mathematics 1135), Springer-Verlag, Berlin, 1985, pp. 9-51.

15. G. V. Chudnovsky, The Thue-Siegel-Roth theorem for values of algebraic functions, Proc. Japan Acad. *59* (1983), 281-284.

16. G. V. Chudnovsky, *The Inverse Scattering Problem and Its Applications to Arithmetic, Algebra and Transcendental Numbers* (Lecture Notes in Physics 120), Springer-Verlag, Berlin, 1980, pp. 103-150.

17. G. V. Chudnovsky, Bäcklund transformations and deformations of linear differential equations with applications to diophantine approximations, in *Problems of Mathematical Physics* (Festschrift in honor of F. Gürsey), Eds. I. Bars, A. Chodas, and C.-H. Tse, Plenum Press, New York, 1984, pp. 201-220.

18. D. V. Chudnovsky and G. V. Chudnovsky, *Padé and Rational Approximations to Systems of Functions and Their Arithmetic Applications* (Lecture Notes in Mathematics 1052), Springer-Verlag, Berlin, 1980, pp. 37-84.

19. G. V. Chudnovsky, Number theoretical applications of polynomials with rational coefficients defined by extremality conditions, in *Arithmetic and Geometry* (Festschrift in honor of I. R. Shafarevich) (Progress in Mathematics 35), Birkhäuser Verlag, 1983, pp. 67-107.

20. L. Pochhammer, Ueber ein Integral mit doppeltem Umlauf, Ueber eine Classe von Integralen mit geschlossener Integrationscurve, Math. Ann. *35* (1889), 470-494; Math. Ann. *37* (1890), 500-511.

21. P. Deligne and G. D. Mostow, Monodromy of hypergeometric functions and non-lattice integral monodromy, Publ. Math. *63* (1986), 5-89.

22. Ch. Hermite, Sur la fonction exponentielle, C. R. Acad. Sci. Paris (1873), 18-24, 74-79, 226-233, 385-293; *Oeuvres*, Vol. III, Gauthier-Villars, Paris, 1912, pp. 150-179.

23. B. Riemann, *Oeuvres Mathematiques*, Albert Blanchard, Paris, 1968.

24. M. V. Fedorjuck, *Steepest Descent Methods*, Nauka, Moscow, 1977 (Russian).

25. S. Ramanujan, Modular equations and approximations to π, Quart. J. Math. *44* (1914), 350-372.

26. G. H. Hardy, *Ramanujan*, 3rd ed., Chelsea, New York, 1978.

27. Ch. Hermite, Sur la théorie des équations modulaires, C. R. Acad. Sci. Paris *48* (1859), *49* (1859); *Oeuvres*, Vol. II, Gauthier-Villars, Paris, pp. 38-42.

28. H. Stark, A transcendence theorem for class number problems I, II, Ann. of Math. *94* (1971), 153-173; *96* (1972), 174-209.

29. D. Shanks, Dihedral quartic approximations and series for π, J. Number Theory *14* (1982), 397-423.

30. A. Weil, *Elliptic Functions According to Eisenstein and Kronecker*, Springer-Verlag, Berlin, 1976.

31. S. Lichtenbaum, On p-adic L-functions associated with elliptic curves, Invent. Math. *56* (1980), 19-55.

32. R. Fricke, *Die Elliptischen Funktionen und ihre Anwendungen*, Vol. I, B. G. Teubner, Stuttgart, 1916.

33. C. L. Siegel, Aum beweise des Starkschen Satzes, Invent. Math. *5* (1968), 180-191.

34. K. Mahler, Perfect systems, Compositio Math. *19* (1968), 99-166.

35. A. L. Galočhkin, Lower bounds of polynomials in the values of a certain class of analytic functions, Math. USSR-Sb. *95* (1974), 396-417.

36. E. Bombieri, On G-functions, in *Recent Progress in Analytic Number Theory*, Vol. 2, ed. H. Halberstam and C. Hooley, Academic Press, New York, 1981, pp. 1-67.

37. G. V. Chudnovsky, Measures of irrationality, transcendence and algebraic independence, Recent progress, in *Journées Arithmétiques*, ed. J. V. Armitage, Cambridge University Press, Cambridge, 1982, pp. 11-82.

38. D. V. Chudnovsky and G. V. Chudnovsky, *Applications of Padé Approximations to the Grothendieck Conjecture on Linear Differential Equations* (Lecture Notes in Mathematics 1135), Springer-Verlag, Berlin, 1985, pp. 52-100.
39. B. Dwork, *Arithmetic Theory of Differential Equations* (Symposia Mathematica 24), Academic Press, New York, 1981, pp. 225-243.
40. G. V. Chudnovsky, On the method of Thue-Siegel, Ann. of Math. *117* (1983), 325-382.
41. H. Padé, *Oeuvres*, Albert Blanchard, Paris, 1984.
42. G. Szegö, *Orthogonal Polynomials*, American Mathematical Society, Providence, R.I., 1978.
43. D. V. Chudnovsky and G. V. Chudnovsky, Backlünd transformations for linear differential equations and Padé approximations, I. J. Math. Pures Appl. (Paris) *61* (1982), 1-16.
44. E. R. Kolchin, Rational approximations to solutions of algebraic differential equations, Proc. Amer. Math. Soc. *10* (1959), 238-244.
45. D. V. Chudnovsky and G. V. Chudnovsky, Rational approximations to linear forms of exponentials and binomials, Proc. Nat. Acad. Sci. U.S.A. *80* (1983), 3138-3141.
46. D. V. Chudnovsky and G. V. Chudnovsky, The Wronskian formalism for linear differential equations and Padé approximations, Adv. in Math. *53* (1984), 28-54.
47. W. N. Schmidt, Thue's equation over function fields, J. Austral. Math. Soc. Ser. A *25* (1978), 385-427.
48. R. C. Mason, The hyperelliptic equation over function fields, Math. Proc. Cambridge, Philos. Soc. *93* (1983), 219-230.
49. Yu. I. Manin, Rational points on algebraic curves over finite fields, Amer. Math. Soc. Transl. (2) *50* (1966), 189-234.
50. E. L. Ince, *Ordinary Differential Equations*, Dover, New York, 1956.
51. C. Runge, Über ganzzahlige Lösungen von Gleichungen zwischen zwei Varänderlichen, J. Reine Angew. Math. *100* (1887), 425-435.
52. B. N. Delone and D. K. Faddeev, *The Theory of Irrationalities of the Third Degree*, American Mathematical Society, Providence, R.I., 1964.
53. M. Hall, The diophantine equation $x^3 - y^2 = k$, in *Computers in Number Theory*, ed. A. O. L. Atkin and B. J. Birch, Academic Press, New York, 1971, pp. 173-198.
54. A. Thue, Über Annäherungswerte algebraischer Zahlen, J. Reine Angew. Math. *135* (1909), 284-305.
55. K. Mahler, Zur Approximation algebraischer Zahlen, I, Über den grössen Primteiler binarer Formen, Math. Ann. *107* (1933), 691-730.
56. A. O. Gelfond, *Transcendental and Algebraic Numbers*, Gostechizdat, Moscow, 1952 (English translation, Dover, New York, 1960).
57. A. Baker, Rational approximation to $\sqrt[3]{2}$ and other algebraic numbers, Quart. J. Math. Oxford *15* (1964), 375-383.
58. E. Bombieri, On the Thue-Siegel-Dyson theorem, Acta Math. *184* (1982), 255-296.

59. D. V. Chudnovsky, *Topological and Algebraic Structure of Linear Problems Associated with Completely Integrable Systems* (Lecture Notes in Physics 180), Springer-Verlag, Berlin, 1983, pp. 65-90.
60. D. V. Chudnovsky, G. V. Chudnovsky, and M. Tabor, Painlevé property of multicomponent isospectral deformation equations, Phys. Lett. A 97 (1983), 268-274.

Computer Algebra and the Hilbert Modular Equation

HARVEY COHN   City College of the City University of New York, New York, New York

1. INTRODUCTION

This brief note can serve only as an indication of a fruitful area for further work using computer algebra for the discovery of Hilbert modular equations. To cite a few examples, MACSYMA has been used successfully by Kaltofen and Yui [5] for the (Weber) modular equation, and REDUCE has been used by Nagaoka [6] for (Hilbert) modular identities. The author has used computers (but no formal algebra) to find the first known Hilbert modular equation for the simplest case [1]. These results are discussed below in the hope that future work will be seen to be feasible.

For any further construction of Hilbert modular equations, the computations are so complicated even to plan that the aid of a suitable computer algebra is no longer a luxury. At the very least, we require the elemental "reversion" of power series, that is, the transformation of a power series in (say) x and y into a power series in $f(x,y)$ and $g(x,y)$, where f and g are bianalytically equivalent to x and y at the origin. This is a necessary method of establishing polynomial identities in functions known only through power series with variable accuracy. Clearly, by sufficient repetition, the process should develop into an "expert system" as more and more types of modular equations are developed.

The research was supported by National Science Foundation Grant MCS-8201717.

2. THE WEBER MODULAR EQUATION

To review the fundamental definitions briefly, the Weber modular function $j(z)$ can be defined in many ways. One convenient way is

$$j(z) = \frac{(1 + 256w)^3}{w}$$
$$w = q\Pi(1 + q^n)^{24} \tag{2.1}$$
$$q = \exp(2\pi i z) \quad (\text{Im } z > 0, \ |q| < 1)$$

Thus $j(z)$ has a Laurent expansion

$$j(z) = \frac{1}{q} + 744 + 196{,}884q + \cdots \tag{2.2}$$

We generally speak of a modular function as having the invariance property

$$z \to Tz \quad T \in PSL(2,Z) \tag{2.3a}$$

[i.e., for $Tz = (az + b)/(cz + d)$, with unimodular integral determinant]. Thus $PSL(2,Z)$ is generated by

$$z \to z + 1 \quad z \to -\frac{1}{z} \tag{2.3b}$$

The importance of the invariance property (2.3a) and (2.3b) is that all such functions (with rational behavior as $q \to 0$) lie in a field generated by $j(z)$.

From these properties it follows that for integral $m > 0$, $j(mz)$ is one of a finite set of conjugates under $PSL(2,Z)$. We take m prime so that the set is conveniently that of $m + 1$ functions, namely,

$$X = \left\{ j(mz), \ j\left(\frac{z}{m}\right), \ j\left(\frac{z+1}{m}\right), \ \cdots, \ j\left(\frac{z+m-1}{m}\right) \right\} \tag{2.4}$$

The symmetric functions of X are polynomials S_i in $j(z)$ with integral coefficients. Thus the $m + 1$ values of X satisfy an algebraic equation (the modular equation)

$$\Phi_m(X, j(z)) = X^{m+1} - S_1 X^m + \cdots + (-1)^{m+1} S_{m+1} = 0 \tag{2.5}$$

For the deepest theoretical results, however, the explicit polynomial Φ_m is not explicitly required; all that is needed is the congruence in $Z[X,Y]$:

$$\Phi_m(X,Y) \equiv (X^m - Y)(Y^m - X) \mod m \tag{2.6}$$

We know additionally that this function $\Phi_m(X,Y)$ is symmetric in X and Y since in (2.4) j(mz) and j(z/m) appear symmetrically. Congruence (2.6) leads to a property useful in class field theory: If X = Y, an algebraic number results which is the value of j(z) for z a quadratic surd. For instance, if $j(\sqrt{-m})$ = Y, then Y = $j(-1/\sqrt{-m})$ [by (2.3b)] and X = $j(m(-1/\sqrt{-m}))$ = Y again. These values (of j for a quadratic surd) are called "singular moduli." They are all algebraic and roots of a modular equation for X = Y. Then this congruence leads to the useful property that when X = Y the common value is an algebraic number whose image in a finite field of characteristic m is invariant under a Frobenius automorphism.

Otherwise, the coefficients are astronomically large, even for small m. Yet for interesting examples of various equations realizing special Galois groups, it is desirable to solve the modular equation for X = Y. This was done in Ref. 5 for m = 7 and 11 using MACSYMA. At the risk of oversimplifying the computation, we might say that the process essentially amounts to expanding the symmetric functions S_k into Laurent series in q and subtracting powers of the series (2.2) for j(z) until a polynomial representation for S_k is achieved.

3. HILBERT MODULAR FUNCTIONS AND EQUATIONS

There is a very imperfect analog (see Ref. 2) of the Weber modular function theory for the Hilbert modular function which makes experimentation all the more desirable. We shall restrict ourselves to the special case of base field $Q(\sqrt{2})$ with integers

$$O = Z[\sqrt{2}] = \{x + y\sqrt{2} : x,y \in Z\} \quad (3.1)$$

The (Hilbert) modular group is now $PSL_2(O)$ and it acts on the product of half-spaces Im z > 0, and Im z' > 0 in the independent variables z and z'. For coefficients in zero, the symbols α and α' denote conjugates. Thus the modular invariance is given by

$$(z,z') \to T(z,z')$$
$$Tz = \frac{\alpha z + \beta}{\gamma z + \delta} \quad Tz' = \frac{\alpha' z + \beta'}{\gamma' z + \delta'} \quad (3.2)$$
$$\alpha = a + b\sqrt{2} \quad \alpha' = a - b\sqrt{2} \quad (a,b \in Z), \text{ etc.}$$

The generating operations of this group are

$$z \to z + 1 \qquad z' \to z' + 1$$
$$z \to -\frac{1}{z} \qquad z' \to -\frac{1}{z'}$$
$$z \to z + \sqrt{2} \qquad z' \to z' - \sqrt{2} \tag{3.3}$$
$$z \to z(1 + \sqrt{2})^2 \qquad z' \to z'(1 - \sqrt{2})^2$$

The function field corresponding to the group $PSL_2(O)$ is not as easily defined as in the earlier case. We must first define a modular form of (even) order 2k, namely, $G(z,z')$, as a holomorphic function for which

$$G(z,z') = G(T(z,z'))[(\gamma z + \delta)(\gamma' z' + \delta')]^{-2k} \tag{3.4}$$

These modular forms are a graded ring generated (see Refs. 3 and 5) by special forms whose definitions we do not pursue further, except to note some expansions in q and q', where

$$q = \exp[\,i(z + z')] \qquad q' = \exp\,i\,\frac{z - z'}{2} \tag{3.5}$$

The basis of the graded ring is given by functions generated with some slight effort as known power series (order indicated by subscript):

$$H_2(z,z') = 1 + 48q\{1,3,1\} + 48q^2\{7,8,15,8,7\} + \cdots$$
$$H_4(z,z') = q\{1,2,-1\} + q^2\{-4,-8,24,-8,-4\} + \cdots \tag{3.6}$$
$$H_6(z,z') = q + q^2\{-2,-16,12,-16,-2\} + \cdots$$

The notation $\{1,3,1\}$ means $q'^{-1} + 3 + q'$, and so on. [It turns out that the forms will have a symmetry with $q' \to 1/q'$ or $(z,z') \to (z',z)$.] It also becomes natural to define the field of modular functions as those representable as a quotient of modular forms of like order. There are no "very natural" generators of this field (of transcendence degree 2), but it is convenient to use

$$X(z,z') = \frac{H_2^2}{H_4} \qquad Y(z,z') = \frac{H_2 H_4}{H_6} \tag{3.7}$$

Here we consider the modular equation with multiplier μ, where for some prime $m \equiv \pm 1 \mod 8$ or $m = 2$,

$$m = x^2 - 2y^2 = \mu\mu' \qquad (\mu = x + y\sqrt{2}) \tag{3.8}$$

If $F(z,z')$ is a modular function, then $F(\mu z, \mu' z')$ is again one of $m + 1$ conjugates

$$W = \left\{ F(\mu z, \mu' z'), F\left(\frac{z}{\mu}, \frac{z'}{\mu'}\right), \ldots, F\left(\frac{z + m - i}{\mu}, \frac{z + m - 1}{\mu'}\right) \right\} \quad (3.9)$$

Then the analogous problem is to construct the equation over X and Y for which the m + 1 roots will be the W. This was done only in a special case m = 2 for $X_0 = X((2 + \sqrt{2})z, (2 - \sqrt{2})z')$ and $Y_0 = Y((2 + \sqrt{2})z, (2 - \sqrt{2})z')$ in terms of $X(z,z')$ and $Y(z,z')$ [1]. The method was to write the symmetric functions of the conjugates W as a quotient of the individual modular forms in the numerators and denominators. Then the remaining problem is to express these modular forms in terms of the basis functions (3.6). The method is to observe the order of the unknown modular forms and set up all possible products of powers $H_2^A H_4^B H_6^C$ of given order 2A + 4B + 6C. Then we find coefficients for these products so that some linear combination represents the numerators and denominators of the symmetric functions of (3.9). In this case only 45 powers were needed (orders 18 and 24 occurred). So any higher case is likely to need more terms than can imaginably be handled without some trial-and-error list procedure.

4. THEORETICAL MOTIVATIONS

The computation of the modular equations has (at least) two points of theoretical interest, one in number theory and the other in algebraic geometry. In number theory there is the "super-theorem" of Weber: Let $F_d(x,y)$ be a principal form of discriminant d (<0),

$$F_d(x,y) = \begin{cases} x^2 - \left(\frac{d}{4}\right) y^2 & d \equiv 0 \mod 4 \\ x^2 + xy - \left(\frac{d-1}{4}\right) y^2 & d \equiv 1 \mod 4 \end{cases} \quad (4.1)$$

Define the so-called "ring class field" (generated by the "singular modulus" shown),

$$K = RCF\{d\} = Q(\sqrt{d}, j(d + \sqrt{d}/2)) \quad (4.2)$$

Then the following property holds for a positive prime p ($\nmid$ 2d):

$$p = F_d(x,y) \Leftrightarrow p \text{ splits } K \quad (4.3)$$

This means that p is represented by the form for integers x, y in Z if and only if all (monic) defining polynomials for (integers in) K are factorable modulo p into linear factors.

There is no analog of similar effectiveness for the next stage of the theory where the forms have coefficients in zero (not Z) and the modular functions (generating "singular moduli") are of Hilbert's (not Weber's) type. There is a rudimentary theory of Hecke [4], but the modular invariants "usually" generate fields of too low a degree [2]. There are very few precise results and there have been a dearth of examples because modular equations are largely unknown.

There is another aspect to this problem involving the equations as algebraic objects. If we look at what we expect in Section 3 (and achieve in Ref. 1 for m = 2), there are two equations of degree m + 1 (recall that m prime $\equiv \pm 1$ mod 8 or m = 2):

$$\Phi_1(X_0,X,Y) = \Phi_2(Y_0,X,Y) = 0 \tag{4.4}$$

Therefore, given X and Y there seem to be $(m + 1)^2$ pairs (X_0,Y_0), which of course is false. Actually, by symmetry,

$$\Phi_1(X,X_0,Y_0) = \Phi_2(Y,X_0,Y_0) = 0 \tag{4.5}$$

This limits the number of solutions. In effect, for given (X,Y), the m + 1 values of (X_0,Y_0) are a variety that probably is not determined by a "complete intersection" (of two equations in X and Y).

The visualization of the workings of the algebraic equations, even for real loci, is probably a matter of investigation for a mixed graphic-algebraic technique.

Added in proof (July 1988): For more recent work, see H. Cohn and J. Deutsch, An explicit modular equation in two variables for $Q(\sqrt{3})$, Math. Comp. *50* (1988), 557-568.

REFERENCES

1. H. Cohn, An explicit modular equation in two variables and Hilbert's twelfth problem, Math. Comp. *38* (1982), 227-236.
2. H. Cohn, Some examples of Weber-Hecke ring class field theory, Math. Ann. *265* (1983), 83-100.
3. K.-B. Gundlach, Die Bestimmung der Funktionen zu einigen Hilbertschen Modulgruppen, J. Reine Angew. Math. *220* (1965), 109-153.
4. E. Hecke, Hohere Modulfunktionen und ihre Anwendungen auf der Zahlentheorie, Math. Ann. *71* (1912), 1-37.
5. E. Kaltofen and N. Yui, Explicit construction of the Hilbert class fields of imaginary quadratic fields with class numbers 7 and 11, (Eurosam 84). Springer Lect. Notes in Computer Sci., Vol. 174 (1984), 310-320.
6. S. Nagaoka, On Hilbert modular forms III, Proc. Japan Acad. Ser. A *59* (1983), 346-348.

Some Cayley Examples

MICHAEL C. SLATTERY   Marquette University, Milwaukee, Wisconsin

1. INTRODUCTION

In my mind, one of the highlights of the conference was the opportunity to see many of the current computer algebra systems being demonstrated. Although I cannot possibly reproduce the hands-on experience available at the conference in this chapter, it would be nice to have some examples of actual terminal sessions with these systems. In particular, we will look at some of the things that can be done with the group theory computer system CAYLEY. The examples presented below are representative of some things that I have done with CAYLEY and certainly do not contain any new or even profound mathematics. They do demonstrate some of the extent to which CAYLEY can be expanded and tailored to fit a person's needs and interests.

Basically, CAYLEY is an interactive computer language that was designed to allow easy access to the growing number of algorithms in computational group theory. CAYLEY allows the user to describe a finite (or finitely presented) group in a number of ways, including generators and relators, permutations, and matrices over a finite field. Once the group is identified, CAYLEY can then perform a number of general group-theoretic calculations such as finding Sylow p-subgroups or calculating centralizers. A number of general descriptions of CAYLEY have appeared in various conference proceedings (e.g., Ref. 1), and John Cannon's notes on the system [2] include a number of very nice examples as well as providing a user's guide to CAYLEY.

2. SEMIDIRECT PRODUCTS

In the study of finite solvable groups, one often hears a group described in terms of a certain normal series rather than any "global" description of the group. Thus G may be a subgroup of some general linear group, but it is more likely to hear it described as an elementary abelian group of order 5^3 with cyclic group of order 6 acting on it in such and such a way. Consequently, in order to describe certain small solvable groups to CAYLEY, it would be convenient to have some sort of semidirect product construction available.

The routine below represents a very rudimentary first step toward such a construction. In particular, given a finitely presented group H and an element x of the automorphism group of H, the procedure "semidirect" constructs the semidirect product of H acted on by the cyclic group generated by x. Despite the limitations of this construction, it still serves to build many interesting small groups.

The basic idea of the construction is very simple—we wish to add a new generator to the presentation of H and then add enough relations to require this new generator to behave like the automorphism x. Unfortunately, CAYLEY does not permit you simply to add a generator to H since this would change the group H. Therefore, we must define a new group G with one more generator than H and attempt to define the proper relations in G. Now we certainly want all the relations of H to hold in G; however, the relations of H are words in the generators of H, while in G we need words in the generators of G. The purpose of the mapping "eta" defined in the procedure is to "translate" these relations. It is interesting to note that eta is not a homomorphism; however, we are only defining it on the generators of H and extending as though it were a homomorphism. One can view eta as a homomorphism from the free group H to the free group G if one is willing to overlook the fact that H is not really a free group (having relations). Despite this theoretical confusion, the map as defined serves our purpose well. Having now created a copy of H in the sub-

Some Cayley Examples

group of G generated by the first k - 1 generators, the routine defines the action of the kth generator g_k to match the action of x. Finally, the order of g_k is defined and we are done.

Here is a listing of the procedure semidirect (phrases enclosed in double quotes are comments):

```
library semi;

"Written by Mike Slattery - April 1984."

procedure semidirect(H,x;G);

"Given a finitely presented group H and an element x of
 the automorphism group of H, this procedure returns G as
 the semidirect product of H and <x>."

  k=ngens(H)+1;
  G:fgrank(k); "Let G be a free group of rank k"
  "The next line defines a map from the generators of H to
   the corresponding generators of G"
  define eta:H to G by images prune(setseq(generators(G)));
  R:=[];
  for each w in relations(H) do
    R:=R join [eta(w)];
  end;
  phix=automorphism(H,x); "phix is the map of H induced by x"
  for i=1 to ngens(H) do
    w:=eta(phix(H.i));
    R:=R join [(G.i^-1)^G.k*w];
  end;
  R:=R join [G.k^order(x)];
  G.relations:R;

end; "semidirect."

finish;
```

As an example of the use of this routine, we will construct a solvable group of order 168. This group can be described as a nonabelian group of order 21 acting on an elementary abelian group of order 8. Alternatively, one can think of the additive group of a field of order 8 acted upon by the multiplicative group of the field, and this in turn acted on by a field automorphism of order 3.

The following log file shows a construction of this group using the procedure semidirect:

VAX CAYLEY   V2.2-27   2-MAY-84 15:53:06   STORAGE   20000

```
>library semi;  "This command loads in the procedure semidirect
                from the library"

>"The next 2 commands define h to be an elementary abelian
> group of order 8"
>h:fgrank(3);
>h.relations:h.1^2,h.2^2,h.3^2,(h.1,h.2),(h.1,h.3),(h.2,h.3);
>
>a=automorphism group(h);
>
>"What I'm doing here is using my semidirect procedure to build
> up a particular solvable group that I want.  The group is a
> non-abelian group of order 21 acting on an elementary abelian
> group of order 8"
>
>print a;

GROUP A OF ORDER 168 = 2^3 * 3 * 7
GENERATORS :
  (4,5)(6,7)
  (4,6)(5,7)
  (1,2)(6,7)
  (1,4)(2,5)
  (1,2,3)(5,6,7)

>"I want to find an element in a of order 7"
>s=sylow(a,7);
>print order(s.1);
     7
>"Now I can build the semi-direct product (of order 56)"
>semidirect(h,s.1;g1); print order(g1);
    56
>
>"Next I want to put the three element on top of this..."
>a=automorphism group(g1);
 ELAPSED: 00:04:58.55  CPU: 0:00:13.54  BUFIO: 2  DIRIO: 0  FAULTS: 2087
>print a;

GROUP A OF ORDER 168 = 2^3 * 3 * 7
GENERATORS :
  (8,13)(9,15)(10,11)(12,14)(16,21)(17,23)(18,19)(20,22)(24,29)(25,31)(26,27)
  (28,30)(32,37)(33,39)(34,35)(36,38)(40,45)(41,47)(42,43)(44,46)(48,51)(49,55)
  (50,52)(53,54)

  (8,15)(9,13)(10,12)(11,14)(16,23)(17,21)(18,20)(19,22)(24,31)(25,29)(26,28)
  (27,30)(32,39)(33,37)(34,36)(35,38)(40,47)(41,45)(42,44)(43,46)(48,55)(49,51)
  (50,53)(52,54)

  (8,11)(9,12)(10,13)(14,15)(16,19)(17,20)(18,21)(22,23)(24,27)(25,28)(26,29)
  (30,31)(32,35)(33,36)(34,37)(38,39)(40,43)(41,44)(42,45)(46,47)(48,52)(49,53)
  (50,51)(54,55)

  (1,2,3,4,5,6,7)(8,13,10,12,15,11,9)(16,21,18,20,23,19,17)(24,29,26,28,31,27,25)
  (32,37,34,36,39,35,33)(40,45,42,44,47,43,41)(48,50,49,52,53,54,55)

  (1,3,7)(2,5,4)(8,27,51)(9,25,52)(10,28,54)(11,30,53)(12,29,55)(13,26,48)
  (14,24,49)(15,31,50)(16,35,46)(17,33,41)(18,36,45)(19,38,40)(20,37,42)
  (21,34,44)(22,32,43)(23,39,47)
```

Some Cayley Examples 93

```
>x=a.5; print order(x); "From the above list we can choose an element of
>                          order 3"
    3
>semidirect(g1,x;g); "and then build the semi-direct product"
>print order(g);
    168
>"Now g is the group I wanted."
>
>"This construction is conceptually much simpler than typing in the
> generators and relators (shown below) but is currently rather slow
> because of the automorphism group calculations."
>
>print g;

GROUP G OF ORDER 168 = 2^3 * 3 * 7
RELATIONS :
 G.1 G.1
 G.2 G.2
 G.3 G.3
 G.1 G.2 G.1 G.2
 G.1 G.3 G.1 G.3
 G.2 G.3 G.2 G.3
 G.4^-1 G.1 G.4 G.3
 G.4^-1 G.2 G.4 G.1 G.3
 G.4^-1 G.3 G.4 G.2
 G.4 G.4 G.4 G.4 G.4 G.4
 G.5^-1 G.1^-1 G.5 G.1
 G.5^-1 G.2^-1 G.5 G.2 G.3
 G.5^-1 G.3^-1 G.5 G.2
 G.5^-1 G.4^-1 G.5 G.3 G.4 G.4
 G.5 G.5 G.5

>
>exit;
```

As mentioned above, this procedure is very limited and suggests some obvious improvements. First, it is often inconvenient to compute the full automorphism group of a group simply to identify a single automorphism. Especially in situations (such as wreath products) where it might be possible to describe the desired automorphism directly, the additional computation is wasteful. Thus one might wish to produce a version of semidirect in which x is simply a map on H (which was probably, but not necessarily, produced by the automorphism function).

The second problem is the fact that one would really like to allow a larger (noncyclic) group to act on H and to allow it to act nonfaithfully. Formally, this is handled by saying that A acts on H if we have a homomorphism from A into the automorphism group of H. Now all of these concepts are available in CAYLEY and one could quite easily write a procedure semidirect(H,A,f;G) which could construct an arbitrary semidirect product. Unfortunately, the difficulty of computing automorphism groups again becomes a serious impediment to this approach.

One other direction that the routine above could be expanded is to try and allow other representations for the group H. For example, it would be nice to build a semidirect product based on a permutation group, and so on.

3. A SMALL p-GROUP EXAMPLE

Recently, as part of an investigation of possible irreducible character degrees of p-groups, a question arose as to whether or not there exists a finite group of order p^5 which has p^2 linear characters and $p^3 - 1$ characters of degree p. Now for those very familiar with small p-groups, the example may be obvious, but this seemed a natural situation in which to use CAYLEY to try and build an example.

Since CAYLEY does not deal with irreducible characters directly, this may seem an inappropriate problem; however, the conditions above can easily be stated in group-theoretic terms: for example, the linear character count is equivalent to requiring that $|G:G'| = p^2$ in our group. If we take Z to be a central subgroup of G of order p such that Z is contained in G', then $Q = G/Z$ must have $Q' = G'/Z$, so Q has exactly p^2 linear characters and $p^2 - 1$ characters of degree p. Such a group had already been constructed, so I knew I could take

$$Q = <a,b,c \mid a^{(p^2)} = b^p = c^p = (a,b) = 1, a^c = a^{(p+1)}*b, b^c = a^p*b>$$

The group G that we want is a central extension of Q, so after a while (much too long), I decided to look at a *darstellungsgruppe* for the group Q. A moment's work with CAYLEY showed that this did in fact provide the desired example. Although the presentation produced by CAYLEY was a little clumsy, it was a simple matter to refine it from that point. We close this section with a log of the CAYLEY session showing this construction.

```
VAX CAYLEY      V2.2-27              8-JUN-84 13:09:01    STORAGE  20000
>"We start by defining Q (for the case p=3)."
>
>Q:free(a,b,c);
>Q.relations:a^9=b^3=c^3=(a,b)=1,a^c=a^4*b,b^c=a^3*b;
>
>print order(Q);
    81
>"Now we use a built-in CAYLEY function to construct G"
>G=darstellungsgruppe(Q);
>print order(G);
   243
>
>"G has the right order - now we check on the character degrees"
>
>H=derived subgroup(G);
>print order(H);
    27
>"Since H has index 9, G must have exactly 9 linear characters"
>
>Z=center(G);
>print order(Z);
     9
```

Some Cayley Examples 95

```
>"Furthermore, since the center of G has index 27, G cannot
> have irreducible characters of degree greater than 3 -
> Thus we have our example."
>print G;

GROUP G OF ORDER 243 = 3^5
RELATIONS :
 G.1 G.1 G.1 G.1 G.1 G.1 G.1 G.1 G.1 G.6^-1
 G.2 G.2 G.2 G.5 G.6
 G.3 G.3 G.3
 G.1^-1 G.2^-1 G.1 G.2 G.4^-1
 G.1 G.1 G.1 G.1 G.2 G.3^-1 G.1^-1 G.3
 G.1 G.1 G.1 G.2 G.3^-1 G.2^-1 G.3
 G.1^-1 G.4^-1 G.1 G.4
 G.2^-1 G.4^-1 G.2 G.4
 G.3^-1 G.4^-1 G.3 G.4
 G.1^-1 G.5^-1 G.1 G.5
 G.2^-1 G.5^-1 G.2 G.5
 G.3^-1 G.5^-1 G.3 G.5
 G.4^-1 G.5^-1 G.4 G.5
 G.1^-1 G.6^-1 G.1 G.6
 G.2^-1 G.6^-1 G.2 G.6
 G.3^-1 G.6^-1 G.3 G.6
 G.4^-1 G.6^-1 G.4 G.6
 G.5^-1 G.6^-1 G.5 G.6

>"Based on the order of G it seems unlikely that all three of
> the new generators G.4, G.5, and G.6 are needed.  In fact
> we can check this with CAYLEY"
>
>K=<G.4,G.5,G.6>;
>print order(K);
     3
>"As suspected, they are just powers of each other"
>
>print order(G.4),order(G.5),order(G.6);
     3   1   1
>"In fact, only G.4 represents a non-trivial element"
>"Thus we can attempt a simpler presentation of G:"
>
>G1:free(u,v,w,z);
>G1.relations:u^9,v^3,(u,v)=z,w^3,u^w=u^4*v, v^w=u^3*v, (u,z), (v,z), (w,z);
>
>print order(G1);
    243
>
>exit;
```

4. CONCLUSION

While CAYLEY has been used (and is very effective) as a teaching tool, the system was designed as a very useful "laboratory" for a research mathematician looking for small examples or attempting to gain insight into the behavior of certain small groups. As well as handling standard computational group theory problems, such as coset enumeration, the features of CAYLEY as an extensible computer language permit a large number of varying problems to be attacked. Furthermore, as new algorithms are developed, many of them are being incorporated into CAYLEY, making it even easier to express a wide range of problems and calculations.

Note added in proof: This article exhibits some aspects of the CAYLEY system as it appeared in 1984. For a more general overview of the system, the reader is encouraged to see [3] and [4]. In particular, [5] contains a more general semidirect product procedure.

REFERENCES

1. J. J. Cannon, *Software Tools for Group Theory*, The Santa Cruz Conference on Finite Groups, Santa Cruz, 1979 (Proceedings of Symposia in Pure Mathematics 37), American Mathematical Society, Providence, R.I., 1980, pp. 495-502.

2. J. J. Cannon, *A Language for Group Theory*, Preprint, University of Sydney.

3. J. J. Cannon, *An Introduction to the Group Theory Language CAYLEY*, Computational group Theory, ed. M. Atkinson, Academic Press, New York, 1984, pp. 145-183.

4. J. J. Cannon, *The Group Theory System CAYLEY*, to appear in Journal of Symbolic Computation.

5. S. Glasby, *Constructing Finite Soluble Groups*, The CAYLEY Bulletin, No. 3, Dept. of Pure Mathematics, University of Sydney, 1987, pp. 106-112.

Physics, Ramanujan, and Computer Algebra

GEORGE E. ANDREWS   Pennsylvania State University, University Park, Pennsylvania

1. INTRODUCTION

The title of this chapter suggests grand scope—to say the least. Actually, our real goals here are somewhat more modest. We shall examine three problems very much in the domain of Ramanujan's interests [4,13]. Our object will be to tackle these problems using computer algebra as an essential tool in our work.

The first problem is that of "discovering" the Rogers-Ramanujan identities following Schur's approach [16]. It is this problem and our computer algebra solution of it that fits into physics. More precisely, Schur's approach was shown to be quite helpful in the work [5] deriving from R. J. Baxter's solution of the hard hexagon model in statistical mechanics. Recently, an infinite family of models generalizing the hard hexagon model was solved [8]. Actual calculation of the various partition functions required proceeded precisely along the lines we present in Section 2.

In Section 3 we briefly sketch how the structure of Bailey chains is ideally suited for study via computer algebra. As a prototype of this work we provide a proof of the Rogers-Ramanujan identities quite different from that in Section 2.

Finally, we present a problem in partitions (Göllnitz's "mod 6" theorem) where the computer is not only useful in discovery but is also immensely helpful in our proof. The proof we present is new and conceptually quite simple; but the algebra needed in its execution is monstrous, to say the least.

I should emphasize that all the computer algebra done in connection with this work was on IBM's SCRATCHPAD package; this is the ancestor of IBM's SCRATCHPAD II, which recently went on-line.

2. THE ROGERS-RAMANUJAN IDENTITIES AND THE HARD HEXAGON MODEL

In 1980, R. J. Baxter solved the hard hexagon model in statistical mechanics; a full account is given by Baxter in Ref. 10. We shall not describe much about his solution except to consider the following question which arose in his considerations of regime I.

Let polynomials D_n (in q) be defined by $D_0 = 1$, $D_1 = 1 + q$, and for $n \geq 2$,

$$D_n = D_{n-1} + q^n D_{n-2} \qquad (2.1)$$

Prove that

$$\lim_{n \to \infty} D_n = \prod_{n=1}^{\infty} \frac{1}{(1 - q^{5n-1})(1 - q^{5n-4})} \qquad (2.2)$$

We are going to show how SCRATCHPAD can be used to provide a conjectured representation of the D_n which leads directly to (2.2). As is often the case in this subject, once we have the right conjecture, its proof goes fairly quickly.

We should mention in passing that the problem posed in (2.2) is the version of the first Rogers-Ramanujan identity that was solved by Schur [16]. Indeed, Schur's solution is quite ingenious and a careful explanation of how Schur's solution was constructed has been given by Rademacher [14]. With much less ingenuity than Schur possessed, we are led directly to his solution via SCRATCHPAD.

The philosophy of our approach is simple. We shall try to write D_n as a sum of Gaussian polynomials. The Gaussian polynomials $\begin{bmatrix} n \\ m \end{bmatrix}$ are defined by

$$\begin{bmatrix} n \\ m \end{bmatrix} = \begin{cases} \dfrac{(1 - q^n)(1 - q^{n-1}) \cdots (1 - q^{n-m+1})}{(1 - q^m)(1 - q^{m-1}) \cdots (1 - q)} & n \geq 0, \; m \geq 0 \\ 0 & n \geq 0, \; m < 0 \end{cases} \qquad (2.3)$$

They are simply encoded in a symbolic algebra package in several ways. For example, their generating function $H_n(x) = \Sigma \begin{bmatrix} n \\ m \end{bmatrix} x^m$ works well in SCRATCHPAD:

$$H_0(x) = 1 \tag{2.4}$$

$$H_n(x) = H_{n-1}(xq) + xH_{n-1}(x) \quad n \text{ in } (1, 2, \ldots) \tag{2.5}$$

$$\begin{bmatrix} n \\ m \end{bmatrix} = \text{coefficient of } x^m \text{ in } H_n(x) \tag{2.6}$$

So we encode the D_n from (2.2) and the $\begin{bmatrix} n \\ m \end{bmatrix}$ from (2.6). We first ask for several D_n:

$$D_0: 1 \tag{2.7}$$

$$D_1: Q + 1 \tag{2.8}$$

$$D_2: Q^2 + Q + 1 \tag{2.9}$$

$$D_3: Q^4 + Q^3 + Q^2 + Q + 1 \tag{2.10}$$

$$D_4: Q^6 + Q^5 + 2Q^4 + Q^3 + Q^2 + Q + 1 \tag{2.11}$$

$$D_5: Q^9 + Q^8 + Q^7 + 2Q^6 + 2Q^5 + 2Q^4 + Q^3 + Q^2 + Q + 1 \tag{2.12}$$

$$D_6: Q^{12} + Q^{11} + 2Q^{10} + 2Q^9 + 2Q^8 + 2Q^7 + 3Q^6 + 2Q^5 + 2Q^4$$
$$+ Q^3 + Q^2 + Q + 1 \tag{2.13}$$

$$D_7: Q^{16} + Q^{15} + Q^{14} + 2Q^{13} + 3Q^{12} + 3Q^{11} + 3Q^{10} + 3Q^9$$
$$+ 3Q^8 + 3Q^7 + 3Q^6 + 2Q^5 + 2Q^4 + Q^3 + Q^2 + Q + 1 \tag{2.14}$$

$$D_8: Q^{20} + Q^{19} + 2Q^{18} + 2Q^{17} + 3Q^{16} + 3Q^{15} + 4Q^{14} + 4Q^{13}$$
$$+ 5Q^{12} + 4Q^{11} + 4Q^{10} + 4Q^9 + 4Q^8 + 3Q^7 + 3Q^6 + 2Q^5$$
$$+ 2Q^5 + 2Q^4 + Q^3 + Q^2 + Q + 1 \tag{2.15}$$

$$D_9: Q^{25} + Q^{24} + Q^{23} + 2Q^{22} + 3Q^{21} + 4Q^{20} + 4Q^{19} + 5Q^{18}$$
$$+ 5Q^{17} + 6Q^{16} + 6Q^{15} + 6Q^{14} + 6Q^{13} + 6Q^{12} + 5Q^{11}$$
$$+ 5Q^{10} + 5Q^9 + 4Q^8 + 3Q^7 + 3Q^6 + 2Q^5 + 2Q^4 + Q^3$$
$$+ Q^2 + Q + 1 \tag{2.16}$$

We now make some simple observations. First, D_{2n} appears always to be of degree $n^2 + n$ and D_{2n-1} of degree n^2. Furthermore, since $\begin{bmatrix} n \\ m \end{bmatrix}$ is a polynomial of degree $m(n - m)$, we see that $\begin{bmatrix} 2n + 1 \\ n \end{bmatrix}$ is of degree $n^2 + n$, and $\begin{bmatrix} 2n \\ n \end{bmatrix}$ is of degree n^2. *Hence let us use* $\begin{bmatrix} 2n+1 \\ n \end{bmatrix}$ *as an approximation to* D_{2n} *and* $\begin{bmatrix} 2n \\ n \end{bmatrix}$ *as an approximation to* D_{2n-1}.

We next ask SCRATCHPAD for the values of $D_{2n-1} - \begin{bmatrix} 2n \\ n \end{bmatrix}$ with n in (1,2,3,4,5). The response is

$$0 \tag{2.17}$$

$$(-Q^2) \tag{2.18}$$

$$-Q^7 - Q^6 - Q^5 - Q^4 - 2Q^3 - Q^2 \tag{2.19}$$

$$-Q^{14} - Q^{13} - 2Q^{12} - 2Q^{11} - 4Q^{10} - 4Q^9 - 5Q^8 - 4Q^7$$
$$- 4Q^6 - 3Q^5 - 3Q^4 - 2Q^3 - Q^2 \tag{2.20}$$

$$-Q^{23} - Q^{22} - 2Q^{21} - 3Q^{20} - 5Q^{19} - 6Q^{18} - 9Q^{17} - 10Q^{18}$$
$$- 12Q^{15} - 13Q^{14} - 14Q^{13} - 14Q^{12} - 14Q^{11} - 13Q^{10}$$
$$- 11Q^9 - 10Q^8 - 8Q^7 - 6Q^6 - 5Q^5 - 3Q^4 - 2Q^3 - Q^2 \tag{2.21}$$

For each $n \geq 2$ it appears that the expressions in q are q^2 times polynomials of degree $n^2 - 4 = (n-2)(n+2)$. This suggests that for our next approximation to D_{2n-1} we use $\begin{bmatrix} 2n \\ n \end{bmatrix} - q^2 \begin{bmatrix} 2n \\ n+2 \end{bmatrix}$.

In the same manner we ask SCRATCHPAD for the values of $D_{2n} - \begin{bmatrix} 2n+1 \\ n \end{bmatrix}$ with n in (0,1,2,3,4):

$$0 \tag{2.22}$$

$$0 \tag{2.23}$$

$$-Q^3 - Q^2 \tag{2.24}$$

$$-Q^9 - 2Q^8 - 2Q^7 - 2Q^6 - 2Q^5 - 2Q^4 - 2Q^3 - Q^2 \tag{2.25}$$

$$-Q^{17} - 2Q^{16} - 3Q^{15} - 4Q^{14} - 5Q^{13} - 6Q^{12} - 7Q^{11} - 8Q^{10}$$
$$- 7Q^9 - 7Q^8 - 6Q^7 - 5Q^6 - 4Q^5 - 3Q^4 - 2Q^3 - Q^2 \tag{2.26}$$

For each $n \geq 2$ it appears that these expressions in q are q^2 times polynomials of degree $n^2 + n - 5 = (n-2)(n+3) + 1$. This suggests that for our next approximation to D_{2n} we use $\begin{bmatrix} 2n+1 \\ n \end{bmatrix} - q^2 \begin{bmatrix} 2n+1 \\ n+3 \end{bmatrix}$.

We now ask SCRATCHPAD for $D_{2n-1} - \begin{bmatrix} 2n \\ n \end{bmatrix} + q^2 \begin{bmatrix} 2n \\ n+2 \end{bmatrix}$, with n in (1,2,3,4,5):

$$0 \tag{2.27}$$

$$0 \tag{2.28}$$

$$(-Q^3) \tag{2.29}$$

$$-Q^{10} - Q^9 - Q^8 - Q^7 - Q^6 - Q^5 - Q^4 - Q^3 \tag{2.30}$$

$$-Q^{19} - Q^{18} - 2Q^{17} - 2Q^{16} - 3Q^{15} - 3Q^{14} - 4Q^{13} - 4Q^{12}$$
$$- 4Q^{11} - 4Q^{10} - 3Q^9 - 3Q^8 - 3Q^7 - 2Q^6 - 2Q^5 - Q^4 - Q^3 \tag{2.31}$$

We also want $D_{2n} - \begin{bmatrix} 2n+1 \\ n \end{bmatrix} + q^2 \begin{bmatrix} 2n+1 \\ n+3 \end{bmatrix}$ for n in (0,1,2,3,4):

$$0 \tag{2.32}$$

$$0 \tag{2.33}$$

$$(-Q^3) \tag{2.34}$$

$$-Q^9 - Q^8 - Q^7 - Q^6 - Q^5 - Q^4 - Q^3 \tag{2.35}$$

$$-Q^{17} - Q^{16} - 2Q^{15} - 2Q^{14} - 3Q^{13} - 3Q^{12} - 4Q^{11} - 4Q^{10}$$
$$- 3Q^9 - 3Q^8 - 3Q^7 - 2Q^6 - 2Q^5 - Q^4 - Q^3 \tag{2.36}$$

Following the same procedure as before, we are led to

$$\begin{bmatrix} 2n+1 \\ n \end{bmatrix} - q^2 \begin{bmatrix} 2n+1 \\ n+3 \end{bmatrix} - q^3 \begin{bmatrix} 2n+1 \\ n-2 \end{bmatrix} \tag{2.37}$$

as a new approximation for D_{2n}, and to

$$\begin{bmatrix} 2n \\ n \end{bmatrix} - q^2 \begin{bmatrix} 2n \\ n+2 \end{bmatrix} - q^3 \begin{bmatrix} 2n \\ n-3 \end{bmatrix} \tag{2.38}$$

as a new approximation for D_{2n-1}.

It should be clear by now that with a symbolic algebra package such as SCRATCHPAD, this process can be carried on indefinitely, and the inevitable conclusion is the following conjecture (found and proved by Schur [16]):

$$D_n = \sum_{\lambda=-\infty}^{\infty} (-1)^\lambda q^{\lambda(5\lambda+1)/2} \begin{bmatrix} n+1 \\ \left[\dfrac{n+1-5\lambda}{2}\right] \end{bmatrix} \tag{2.39}$$

where $\begin{bmatrix} a \\ b \end{bmatrix}$ is the Gaussian polynomial and [a] is the greatest integer in a. Once you have conjectured (2.39), the proof is quite straightforward (see, e.g., Ref. 1).

It should be reemphasized that this approach to Schur's treatment of the Rogers-Ramanujan identities was that followed to provide the explicit expressions for the Rogers-Ramanujan identities arising in R. J. Baxter's solution of the hard hexagon model (Ref. 5), and that this approach appeared to be the only effective one available in the extensive generalizations of the hard hexagon model given in Ref. 8.

3. THE ROGERS-BAILEY APPROACH TO THE ROGERS-RAMANUJAN IDENTITIES

The use of SCRATCHPAD that we now consider relies on Rogers's [15] original approach to the Rogers-Ramanujan identities. Actually, the devices that are most effective were originated in Bailey's [9] extension of Rogers's ideas. Subsequently, it has been shown how each formula originally discovered by Rogers can be embedded in an infinite family of such identities (Ref. 6), and Bailey's lemma has also been applied successfully to the fifth- and seventh-order mock theta functions, with the essential aid of SCRATCHPAD [7].

The approach is as follows: If for each $n \geq 0$,

$$\beta_n = \sum_{j=0}^{n} \frac{\alpha_j}{(q)_{n-j}(q)_{n+j}} \tag{3.1}$$

then (assuming that suitable convergence conditions are met, usually guaranteed by $|q| < 1$)

$$\sum_{n=0}^{\infty} q^{n^2} \beta_n = \frac{1}{(q)_\infty} \sum_{n=0}^{\infty} q^{n^2} \alpha_n \tag{3.2}$$

Furthermore, the sequences

$$\alpha_n' = q^{n^2} \alpha_n \tag{3.3}$$

and

$$\beta_n' = \sum_{j=0}^{n} \frac{q^{j^2} \beta_j}{(q)_{n-j}} \tag{3.4}$$

also fulfill (3.2). In (3.1)-(3.4) we have utilized the classical notation [18; p. 89]

$$(q)_n = (1-q)(1-q^2)\cdots(1-q^n) \tag{3.5}$$

and

$$(q)_\infty = \prod_{n=1}^{\infty} (1-q^n) \tag{3.6}$$

The equations (3.1)-(3.4) constitute simplified parts of what has been called Bailey's lemma [6]; certainly in this form these relationships were mostly known to L. J. Rogers [15].

Now let us turn to the formulation of the first Rogers-Ramanujan identity which was given by Rogers.

$$\sum_{n=0}^{\infty} \frac{q^{n^2}}{(q)_n} = \prod_{n=1}^{\infty} \frac{1}{(1-q^{5n-4})(1-q^{5n-1})}$$

$$= \frac{1}{(q)_{\infty}}\left[1 + \sum_{n=1}^{\infty} (-1)^n q^{n(5n-1)/2}(1+q^n)\right] \quad (3.7)$$

The identity of the second and third members of (3.7) is an easy consequence of Jacobi's triple product identity [3, Cor. 2.9, p. 22]. Let us then consider the left-hand side of (3.7) as a possible instance of the left-hand side of (3.2). Thus we want to examine the α_j's defined by (3.1) when $\beta_n = 1/(q)_n$. Hence with $\alpha_0 = \beta_0 = 1$, we want, for $n \geq 1$,

$$\alpha_n = (q)_{2n}\left[\frac{1}{(q)_n} - \sum_{j=0}^{n-1} \frac{\alpha_j}{(q)_{n-j}(q)_{n+j}}\right]$$

$$= \prod_{j=n+1}^{2n}(1-q^j) - \sum_{j=0}^{n-1}\begin{bmatrix}2n \\ n-j\end{bmatrix}\alpha_j$$

We encode this recurrence in SCRATCHPAD and ask for α_n with n in (1,2,3,4,5). This request quickly yields

$$\alpha_1: -q^2 - q \quad (3.9)$$
$$\alpha_2: q^7 + q^5 \quad (3.10)$$
$$\alpha_3: -q^{15} - q^{12} \quad (3.11)$$
$$\alpha_4: q^{25} + q^{22} \quad (3.12)$$
$$\alpha_5: -q^{40} - q^{35} \quad (3.13)$$

and one immediately sees that very likely

$$\alpha_n = (-1)^n q^{n(3n-1)/2}(1+q^n) \quad (3.14)$$

Furthermore, this conjecture causes the right-hand side of (3.2) to coincide with the third member of (3.7). Thus SCRATCHPAD has strongly suggested to us that (3.7) is valid given (3.2). However, we still have before us the problem of actually proving our conjecture that (3.1) is fulfilled by $\beta_n = 1/(q)_n$ and $\alpha_n = (-1)^n q^{n(3n-1)/2}(1 + q^n)$.

Surprisingly enough, we can invoke (3.4) and (3.5) to help us prove our results, again relying on SCRATCHPAD for guidance. Let us consider in (3.3) that $\alpha'_n = (-1)^n q^{n(3n-1)/2}(1 + q^n)$. Then in this case the new α_n is $(-1)^n q^{n(n-1)/2}(1 + q^n)$. We want to show that $\beta'_n = (1/(q)_n$; so we ask SCRATCHPAD to tell us the first few values of β_n corresponding to $\alpha_n = (-1)^n q^{n(n-1)/2}(1 + q^n)$. The response is

$\beta_0: 1$ (3.15)

$\beta_1: 0$ (3.16)

$\beta_2: 0$ (3.17)

$\beta_3: 0$ (3.18)

$\beta_4: 0$ (3.19)

Hence a new conjecture for our new pair of sequences is given by

$$1 + \sum_{j=1}^{n} \begin{bmatrix} 2n \\ n - j \end{bmatrix} (-1)^j q^{j(j-1)/2}(1 + q^j) = \begin{cases} 1 & n = 0 \\ 0 & n > 0 \end{cases} \quad (3.20)$$

Now (3.20) is quite easy to prove. It is trivial for $n = 0$, and for $n > 0$,

$$1 + \sum_{j=1}^{n} \begin{bmatrix} 2n \\ n - j \end{bmatrix} (-1)^j q^{j(j-1)/2}(1 + q^j)$$

$$= \sum_{j=-n}^{n} \begin{bmatrix} 2n \\ n + j \end{bmatrix} (-1)^j q^{j(j-1)/2}$$

$$= (-1)^n q^{n(n+1)/2} \sum_{j=0}^{n} \begin{bmatrix} 2n \\ j \end{bmatrix} (-1)^j q^{-nj+j(j-1)/2}$$

$$= (-1)^n q^{n(n+1)/2} \prod_{m=0}^{2n-1} (1 - q^{-n+m}) \quad ([3; \text{ e.g., } (3.3.6), \text{ p. 36}])$$

$$= 0$$

Furthermore, all we have used in proving (3.20) is the well-known q-binomial theorem [3, e.g., (3.3.6), p. 36].

Now, as simple as (3.20) is, it is adequate to prove the first Rogers-Ramanujan identity. We begin by observing that (3.20) implies that (3.1) holds when $\alpha_0 = 1$, $\alpha_n = (-1)^n q^{n(n-1)/2}(1 + q^n)$ otherwise, and $\beta_n = 1$ if n = 0 and 0 otherwise. Hence by (3.3) and (3.4), we see that (3.2) is satisfied by $\alpha_0' = 1$, $\alpha_n' = (-1)^n q^{n(3n-1)/2}(1 + q^n)$ otherwise and $\beta_n' = 1/(q)_n$; that is, (3.7) has been verified.

As we mentioned earlier, this process of producing sequences of pairs of sequences α_n, β_n has been explored extensively in Ref. 6 and has also been applied to the fifth- and seventh-order mock theta functions [7]. In the latter instance SCRATCHPAD was essential in the production of sequences α_n, β_n which could be proved to fulfill (3.1).

4. GÖLLNITZ'S THEOREM

We finally consider how SCRATCHPAD can be utilized when the algebraic manipulation required in a proof becomes completely unmanageable. For our illustration we provide a new proof of H. Göllnitz's result [12, Th. 4.1]:

THEOREM 4.1 Let A(n) denote the number of partitions of n into parts $\equiv$ 2, 5, 11 (mod 12). Let B(n) denote the number of partitions of n into distinct parts $\equiv$ 2,4,5 (mod 6). Let C(n) denote the number of partitions of n of the form $n = b_1 + b_2 + \cdots + b_s$, where $b_i - b_{i+1} \geq 6$ with strict inequality if $b_i \equiv 0,1,3$ (mod 6) and where $b_s \neq 1,3$. Then A(n) = B(n) = C(n).

The fact that A(n) = B(n) follows directly from a simple argument involving infinite products [2, eq. (1.1)]. The real problem is to prove B(n) = C(n). There are two known proofs of this fact. The first, by Göllnitz [12], is patterned after Gleissberg's [11] proof of Schur's theorem [17]. The disadvantage of this proof is easy to see upon inspection; for example, equation (4.23) of Ref. 12 occupies over a page and a half of Crelle's Journal. The other known proof relies on an intricate algebraic argument related to a q-analog of Tchebychev inversion [2, Th. 2].

We shall provide a new proof here which is again algebraically tedious but conceptually simple. However, this time, we let SCRATCHPAD absorb all the technical problems. Actually, we shall prove a refinement of B(n) = C(n). Namely:

THEOREM 4.2 Let B(n,m) denote the number of partitions of n into m distinct parts $\equiv$ 2,4,5 (mod 6). Let C(n,m) denote the number of partitions of n of the form $n = b_1 + b_2 + \cdots + b_s$, where $b_i - b_{i+1} \geq 6$ with strict inequality if $b_i \equiv 0,1,3$ (mod 6), where $b_s \neq 1,3$ and m is the number of $b_i \equiv 2,4,5$ (mod 6) plus *twice* the number of $b_i \equiv 0,1,3$ (mod 6). Then B(n,m) = C(n,m) for each n and m.

Proof: We begin by defining a sequence of polynomial generating functions related to $C(n,m)$. Let $C(n,m;N)$ denote the number of partitions of the type enumerated by $C(n,m)$, with the added condition that the largest part is $\leq N$. Now define

$$d_N(t) = d_N(t,q) = \sum_{n,m \geq 0} C(n,m;N) t^m q^n \tag{4.1}$$

Then this sequence is defined by $d_0(t) = d_1(t) = 1$, $d_2(t) = d_3(t) = 1 + tq^2$, $d_4 = 1 + tq^2 + tq^4$, $d_5(t) = 1 + tq^2 + tq^4 + tq^5$, and

$$d_{6n-a}(t) = d_{6n-a-1}(t) + tq^{6n-a} d_{6(n-1)-a}(t) \quad n \geq 2, \ a \in (2,4,5) \tag{4.2}$$

$$d_{6n-a}(t) = d_{6n-a-1}(t) + t^2 q^{6n-a} d_{6(n-1)-a-1}(t) \quad n \geq 2, \ a \in (0,1,3) \tag{4.3}$$

On the basis of the cases $n = 1, 2$, it was conjectured about 15 years ago that

$$d_{6n+3}(t) - tq^{6n+2} d_{6n-3}(t) = (1 + tq^2)(1 + tq^4)(1 + tq^5) d_{6n-7}(tq^6) \tag{4.4}$$

[where $d_{-1}(t) = 1$].

If (4.4) is true, then Theorem 4.2 follows easily, because it is clear that

$$d_\infty(t) = \lim_{N \to \infty} d_N(t) = \sum_{n,m \geq 0} C(n,m) t^m q^n \tag{4.5}$$

and by (4.4),

$$d_\infty(t) = (1 + tq^2)(1 + tq^4)(1 + tq^5) d_\infty(tq^6) \tag{4.6}$$

Iteration (4.6) leads immediately to the result that

$$d_\infty(t) = \prod_{n=0}^{\infty} (1 + tq^{2+6n})(1 + tq^{4+6n})(1 + tq^{5+6n})$$

$$= \sum_{n,m \geq 0} B(n,m) t^m q^n \tag{4.7}$$

Comparison of (4.7) with (4.5) yields the desired conclusion that $B(n,m) = C(n,m)$ always holds.

Thus our problem now revolves around proving that our polynomials $d_N(t)$ actually do satisfy (4.4). First, we note that the subsequence $d_{6n-1}(t)$ satisfies a fourth-order recurrence, namely,

$$F(n,t,d_{6n-1}(t),d_{6n-7}(t),d_{6n-13}(t),d_{6n-19}(t),d_{6n-25}(t)) = 0 \quad (4.8)$$

$$(n \geq 3)$$

where

$$\begin{aligned}F(n,t,X,Y,Z,W,V) = &X - [1 + tq^{6n}(q^{-1} + q^{-2} + q^{-4})]Y \\ &- t^2 q^{6n}(q^{-3} + q^{-5} + q^{-6})(1 - q^{6n-6})Z \\ &- t^3 q^{12n-13}[t(q^{-1} + q^{-2} + q^{-4}) + q^{6n-12}]W \\ &- t^6 q^{18n-32} V\end{aligned} \quad (4.9)$$

[and $d_{-1}(t) = 1$, $d_{-7}(t) = 0$]. Now this is a straightforward exercise that can easily be done by hand and can be produced directly from the work on page 40 of Ref. 2. In the same way we can prove that

$$d_{6n+3}(t) - tq^{6n+2} d_{6n-3}(t) = G(n,t) \quad (4.10)$$

where

$$\begin{aligned}G(n,t) = &d_{6n+5}(t) - tq^{6n+2}(1 + q^2 + q^3)d_{6n-1}(t) \\ &+ t^2 q^{12n}(1 + q + q^3)d_{6n-7}(t) - t^3 q^{18n-7} d_{6n-13}(t)\end{aligned} \quad (4.11)$$

To identify the left side and right side of (4.4), it suffices to show that each side satisfies the same fourth-order recurrence and that the identity is valid for a sufficient number of initial values.

The right-hand side of (4.4) obviously satisfies [by (4.8)]

$$F(n - 1, tq^6, d_{6n-7}(tq^6), d_{6n-13}(tq^6), d_{6n-19}(tq^6), d_{6n-25}(tq^6),$$

$$d_{6n-31}(tq^6)) = 0 \quad (n \geq 4) \quad (4.12)$$

Let us formally denote the expression on the left-hand side of (4.8) by $J(n)$. We want in light of (4.9) to show that for $n \geq n_0$,

$$F(n - 1, tq^6, G(n,t), G(n - 1, t), G(n - 2, t),$$

$$G(n - 3, t), G(n - 4, t)) = 0 \quad (4.13)$$

We ask SCRATCHPAD to see if we can write the left-hand side of (4.13) in terms of $J(n)$ by examining the coefficients of $d_{6n+5}(t)$, then $d_{6n-1}(t)$, then $d_{6n-7}(t)$, and so on. Thus

$$F(n - 1, tq^6, G(n,t), G(n - 1, t), G(n - 2, t), G(n - 3, t), G(n - 4, t))$$
$$= J(n + 1) - tq^{6n-6}(q^2 + q^4 + q^5)J(n)$$
$$+ t^2 q^{12n-12}(1 + q + q^3)J(n - 1) - t^3 q^{18n-25} J(n - 2) \qquad (4.14)$$

Now $J(n) = 0$ for $n \geq 3$. Hence (4.13) is valid for $n \geq 5$.

Now we see by (4.12) and (4.13) that both sides of (4.4) satisfy the same fourth-order recurrence starting at $n = 5$. Thus we must verify (4.4) for $n = 1, 2, 3, 4$, and this is easily accomplished by asking SCRATCHPAD to check these cases. This then concludes the proof of (4.4), and with it, Theorems 4.1 and 4.2.

We should remark that in this section the role that SCRATCHPAD played appears minimal. However, that is not really true. Since $d_{6n+3}(t)$ is a polynomial of degree $3n^2 + 5n + 2$ in q, we see that the case $n = 4$ of (4.4) involves polynomials of degree 70 in q. Also, the succinctness of notation in (4.14) masks the hugeness of this expression when considered as a linear combination of $d_{6n+5}(t)$, $d_{6n-1}(t)$, ..., $d_{6n-37}(t)$.

5. CONCLUSION

In this brief survey we have tried to illustrate some of the uses of a symbolic algebra package in studying the types of problems considered by Ramanujan. Greater depth of presentation can be found in several sources: For Section 2, see Ref. 8, and for Section 3, see Refs. 6 and 7.

ACKNOWLEDGMENTS

This work was partially supported by National Science Foundation Grant MCS-8201733. I wish to express my gratitude to Richard Jenks of IBM and David Yun (formerly of IBM) for making it possible to have SCRATCHPAD implemented at the Pennsylvania State University.

REFERENCES

1. G. E. Andrews, A polynomial identity which implies the Rogers-Ramanujan identities, Scripta Math. *28* (1970), 297-305.
2. G. E. Andrews, On a partition theorem of Göllnitz and related formulae, J. Reine Angew. Math. *236* (1969), 37-42.
3. G. E. Andrews, The theory of partitions, in *The Encyclopedia of Mathematics and Its Applications*, Vol. 2, ed. G.-C. Rota, Addison-Wesley, Reading, Mass., 1976.

4. G. E. Andrews, An introduction to Ramanujan's "lost" notebook, Amer. Math. Monthly 86 (1979), 89-108.

5. G. E. Andrews, The hard-hexagon model and Rogers-Ramanujan type identities, Proc. Nat. Acad. Sci. U.S.A. 78 (1981), 5290-5292.

6. G. E. Andrews, Multiple series Rogers-Ramanujan type identities, Pacific J. Math. 114 (1984), 267-283.

7. G. E. Andrews, The fifth and seventh order mock theta functions. Trans. Amer. Math. Soc. 293 (1986), 113-134.

8. G. E. Andrews, R. J. Baxter, and P. J. Forrester, Eight-vertex SOS model and generalized Rogers-Ramanujan-type identities, J. Statist. Phys. 35 (1984), 193-266.

9. W. N. Bailey, Identities of the Rogers-Ramanujan type, Proc. London Math. Soc. (2) 50 (1949), 1-10.

10. R. J. Baxter, *Exactly Solved Models in Statistical Mechanics*, Academic Press, New York, 1982.

11. W. Gleissberg, Über einen Satz von Herrn I. Schur, Math. Z. 28 (1928), 372-382.

12. H. Göllnitz, Partitionen mit Differenzenbedingungen, J Reine Angew. Math. 225 (1967), 154-190.

13. G. H. Hardy, *Ramanujan*, Cambridge University Press, Cambridge, 1940 (Reprinted: Chelsea, New York, 1971).

14. H. Rademacher, *Topics in Analytic Number Theory*, Springer-Verlag, 1973.

15. L. J. Rogers, Second memoir on the expansion of certain infinite products, Proc. London Math. Soc. 25 (1894), 318-343.

16. I. Schur, Ein Beitrag zur additiven Zahlentheorie und zur Theorie der Kettenbrüche, Abh. Preuss. Akad. Wiss. Phys. Math. Kl. (1917), 302-321 (Reprinted in I. Schur, *Gesammelte Abhandlungen*, Vol. 2, Springer-Verlag, Berlin, 1973, pp. 117-136.

17. I. Schur, Zur additiven Zahlentheorie, Abh. Preuss. Akad. Wiss. Phys. Math. Kl. (1926), 488-495 (Reprinted in I. Schur, *Gesammelte Abhandlungen*, Vol. 3, Springer-Verlag, Berlin, 1973, pp. 43-50).

18. L. J. Slater, *Generalized Hypergeometric Functions*, Cambridge University Press, Cambridge, 1966.

POLYPAK: An Algebraic Processor for Computations in Celestial Mechanics

DIETER S. SCHMIDT   University of Cincinnati, Cincinnati, Ohio

1. INTRODUCTION

Celestial mechanics has frequently provided an impetus for the development of new areas of research in mathematics. Therefore, it is not surprising that some of the first programs to do symbolic computations by machine were written to solve problems in this field. Actually, celestial mechanics was predestined for this, as it has a tradition of requiring long algebraic calculations. The best known case is the one of Delaunay [1]. He spent 20 years calculating the position of the moon only to discover in the end that he had not gone far enough in his computations. The actual observations of the moon in the second half of the nineteenth century were already more accurate than his calculations could provide.

To check Delaunay's work and to improve on it, Deprit [2] wrote in the late 1960s his program for the mechanized algebraic operations (MAO) of multiple Fourier series. Such series arise naturally in celestial mechanics and MAO was designed to manipulate truncated versions of these series formally. With action variables I_j and angular variables θ_j ($j = 1, \ldots, n$), the series that occur have the form

$$\Sigma \ A_k \ {\cos \atop \sin} (k_1 \theta_1 + \cdots + k_n \theta_n) \qquad (1.1)$$

The coefficients A_k are power series in the action variables. Summation is over all integer tuples $k = (k_1,\ldots,k_n)$ with the restriction that the first nonzero integer from the left is positive. Such series were named Poisson series by Deprit (not so much because Poisson had contributed to the use of such series but because Poisson's name was often misspelled or incorrectly pronounced in the United States).

An algebraic processor for Poisson series provides at least the following basic operations: Creation and deletion of a series, multiplication of a series with a scalar, addition and multiplication of two Poisson series, and differentiation with respect to one of the variables. Integration is usually required with respect to time where one can assume that the action variables are constant and the angular variables vary linearly with time.

Compared to a general-purpose system such as MACSYMA, REDUCE, or SCRATCHPAD, the type of operations that are required for Poisson series are few in number and not difficult to implement. On the other hand, series that arise in celestial mechanics may have hundreds or even thousands of terms and it becomes important to implement the operations on these series efficiently. Since the general systems have not addressed this specific problem, researchers in celestial mechanics continue to write their own algebraic processors. They achieve efficiency by taking advantage of the available hardware or by writing parts (or all) of the program in assembler language. The disadvantage is that these systems are not portable.

Another difference from the general-purpose systems is that most of the time it suffices to do the calculations in floating-point arithmetic. The coefficients of the power series A_k in (1.1) can be kept as floating-point numbers. The reason is that often a problem depends on a physical parameter which is known with finite precision anyway and that the series are used in a truncated form. It then makes little sense to compute the coefficients exactly when the result is afflicted with a truncation error.

The design and implementation of an algebraic processor is a suitable project for a course on data structures in a class on computer programming. Nevertheless, the successful implementation of such a system to the point where substantial problems can be worked on is still a major undertaking. Time and effort have to be spent to make the system efficient and easy to use.

POLYPAK is our package of computer programs for the formal manipulation of real or complex power series in several variables. The program

originated in a course on PL/I programming that A. Deprit gave at the
University of Cincinnati in 1977. Over the years it has gone through several changes. Like any private programming project, it may never reach a
final form as long as it does not have to be maintained and documented for
other users. On the other hand, the features that give POLYPAK its power
will, in all likelihood, not be changed. This report is a description of
the important features of POLYPAK and it is followed by a summary of the
research projects to which POLYPAK has been applied.

2. DESCRIPTION OF POLYPAK

The series that POLYPAK can manipulate have the form

$$\Sigma \, c_j x_1^{j_1} \cdots x_m^{j_m} \qquad (2.1)$$

Summation is over all integer tuples $j = (j_1, \ldots, j_m)$. The coefficients C_j
can be real or complex numbers either in rational or floating-point format.
On first sight this form appears different from the one for Poisson series,
but introduction of the variables

$$x_e = \exp(i\theta_e) \qquad 1 \le e \le n \qquad (2.2)$$

and the use of the identities

$$\cos \theta_e = \frac{1}{2}(x_e + x_e^{-1}) \qquad (2.3a)$$

$$\sin \theta_e = \frac{1}{2i}(x_e - x_e^{-1}) \qquad (2.3b)$$

show that the two forms are equivalent.

Because of this, certain variables in (2.1) can be selected to be of
the form (2.2) that is with unit modulus. Other variables can be selected
to be purely real, whereas still others can represent general complex variables which can have arbitrary real or imaginary parts. In the last case
another variable may have to designate the conjugate complex value if it
is needed during the computations.

POLYPAK is written for the PL/I optimizing compiler of IBM (release
4.0). This language has a preprocessor. We use it so that for every application the most appropriate version of POLYPAK is compiled. It is done
by setting certain macro variables. They determine how many variables and
of which type are to be used in (2.1) and of what type and precision the

coefficients should be. This information is needed to determine how much space each monomial in (2.1) will require. The set of exponents $(j_1,\ldots, j_m)$ will be stored in an array of halfword integers. The type and precision tell us how much space has to be set aside for each coefficient. The allocation of a fixed amount of space is standard practice with floating-point arithmetic. The user of POLYPAK can select, in the terminology of IBM, between single, double, or extended precision.

Setting aside a fixed amount of space for a rational number poses a certain constraint, as the integers that make up the rational numbers have the tendency to get larger and larger during the computations. This happens even if all precautions are taken to remove any common divisors as soon as possible. When the size of the rational numbers is restricted, an overflow can occur. In this case the change of a single macro variable and recompilation of the entire program allow us to repeat the calculations with floating-point arithmetic. As mentioned earlier, such an approach is normally satisfactory for problems in celestial mechanics. At the moment we allow for three choices for the size of the numerator and denominator that compose a rational number and three different types of arithmetic come with it. A user of POLYPAK can choose full-word binary integers ($< 2^{31}$) or double-word decimal integers ($< 10^{15}$) or double-word binary integers ($< 2^{63}$), where the arithmetic is supplied by assembler subroutines.

A series is generated by adding monomials to it one at a time. Space is provided on "sheets" of a fixed size. These sheets can be identical to the pages of the virtual storage system of the computer except that we also provide a macro variable so that the size of a sheet can be selected to be a fraction of a page. The sheets are managed by POLYPAK. They are given to a series as the need arises and they are reclaimed when a series is deleted.

It will now become apparent why it is more efficient for us to reserve the same amount of space for each monomial. If each monomial requires M bytes (this value includes 4 bytes for two links) and the size of a sheet is N bytes, then $[(N - 4)/M]$ monomials will fit onto one sheet and they will always be located at the same positions relative to the beginning of each sheet. The individual sheets for each series are linked together linearly by their absolute addresses, which accounts for the 4 in the formula above. The end of the linear list is indicated by the absolute address zero. A series itself is given by an address variable

that points to the first sheet. If the series has no nonzero terms, the
pointer variable will hold the address zero and thus no additional space
in the form of a sheet is needed.

The monomials of a series form the nodes of a balanced binary tree.
The exponents determine the ordering within the binary tree; that is, each
array of exponents is treated as a character string and the comparisons
are performed with it. The nodes are linked together by their node number,
which ranges between 0 and 32767. Since a link of 0 indicates the end of a
branch, node 0 can conveniently be used as a control block for the series.
It will tell where is the root of the tree, which is the next available
node, and it contains information about special properties of the series.

Balanced binary trees have the property that at each node the left
and right subtree differ at most by one in their heights (see Knuth [5]).
It ensures that in the worst case a tree with 32767 nodes has at most a
height of 21.

The insertion into balanced binary trees causes a certain amount of
overhead in the form of rebalancing the tree whenever it is needed. This
overhead can be minimized if the nodes are presented for insertion in a
random sequence. In the manner in which we have implemented the algebraic
operations we hope that this will work to our advantage. For example, addition of two series is provided by $Z = Z + C * X$; that is, each term of
the series X is multiplied first by a scalar C and then added to the series
Z. The nodes of X are gone through one by one (in the order in which they
have been inserted into X), irrespective of their order within the binary
tree. First, it is very simple to go through series X in this way, and
second, the series Z may not have to be rebalanced quite as often when the
new terms are added to it in this sequence. This is even more important
for the multiplication of two series as it is one of the most time-consuming operations. Multiplication between two series X and Y is given by

$$Z = Z + C * X * Y \qquad (2.4)$$

where the individual products of X and Y are multiplied with a scalar C
before they are added to Z. For the type of series with which we are working it seems to be best to multiply each term of X with each term of Y in
sequence.

It is possible to introduce shortcuts when the computations are performed with floating-point coefficients. In this case we keep only terms
which are above a threshold in absolute value. Instead of computing all

possible products and discarding those which are too small, we first rearrange the series Y by decreasing size of the absolute value of its coefficient. This can be done without destroying the binary tree structure of the series. The term with the largest coefficient is then in node 1. When we find that a product of a term in X with a term in Y is below our threshold, we know that all following products with that term in X can be ignored. The ordering of a series takes some time, but the overall savings can be substantial. Frequently, no new terms are added to a series after it has participated as an operand in an algebraic operation, and thus it can be kept in the ordered form.

Multiplication of two series is performed frequently. For example, it is used when we raise a series X to the power a/b with a and b integers. Such an operation is available with floating-point coefficients, where we try to compute all terms above our threshold. The same operation is available for all versions of POLYPAK, but in this case one of the real variables has to play the role of a small parameter and we will compute all terms up to a predetermined power in this parameter.

Besides the standard differentiation, POLYPAK provides for the differential operator $D_e = x_e d/dx_e$, which comes up naturally in conjunction with (2.1) when one differentiates with respect to θ_e, which gives $D_e = id/d\theta_e$. Other operations that are available include the Lie derivative of two series and algebraic operations that involve the conjugate complex of a series. It serves no purpose here to enumerate them all, but we like to mention that it is not difficult to introduce additional operations for the series. This can be done in the form of subroutines or directly in the user's program, as the terms of a series are easily accessible. A macro statement generates the required loops, which will traverse all nodes on each sheet.

When a Poisson series has only cosine terms or only sine terms, the corresponding power series has to keep twice as many terms as (2.3a) and (2.3b) show. As such series are encountered, POLYPAK sometimes provides specialized services in creating them. In this case only half of the terms are stored, but before such a series is used, one has to remember to add the conjugate complex of this series. Of course, such an addition is not done explicitly, but special routines that work on these series take it into account automatically. Information in the control block (node 0) distinguishes these series from the others. As the form of an answer provides a first clue if it can be correct, we do not use this feature until a program is debugged and we have to reduce the computing time in order to get the final answer.

Without secondary storage POLYPAK could not deal with the extensive problems for which it was designed. The use of relative links within each binary tree makes each series relocatable. It can be moved to disk and only the links from one sheet to the next have to be updated.

For secondary storage we use the REGIONAL(1) data set of PL/I. It allocates regions with the size of our sheets and each region is known by its region number. To distinguish a region number from an absolute memory address, we negate the region number of the first sheet before we put its value into the pointer variable that represents the series. In this way we know when a series is out on disk. We could let all of our routines decide by themselves when to bring a series into core. We have abandoned this idea because it led to too much movement of series between core and disk. Now it is up to users of POLYPAK to decide when they need space in core and when to bring into core a series from disk.

Besides a temporary secondary storage a permanent version is also available. There, region 0 is set up as a directory so that a series can be referenced by a symbolic name. The name is restricted to four characters in order to get a reasonably sized directory. The last two characters can be decimal digits, so that it is possible to save arrays of series in a limited sense under a symbolic name (i.e., "U1").

The request for any operations by POLYPAK is initiated by a call to the appropriate subroutine. For example, multiplication between two series as stated in (2.4) is requested by CALL MULT(Z,C,X,Y). It is possible to let the preprocessor of PL/I generate these calls from statements such as (2.4). The operation required then has to be an argument of a preprocessor function that we call LET. This approach is not very satisfactory when we would like to parse arbitrary expressions that include operations on series. The reason is that the preprocessor of PL/I does not have access to the symbol table of the compiler and thus can not distinguish between scalars and series. The correct calls to the appropriate subroutines cannot be generated without knowing the type of variables. What is missing is the capability to declare a variable of type series and the possibility of overloading the algebraic operators. These facilities will be available in Ada. For this reason we did not pursue this idea further in the current version of POLYPAK, as we plan to rewrite POLYPAK in Ada as soon as it becomes feasible.

It would also have been more convenient if we could have run POLYPAK in an interactive mode and monitor the progress of a computation. Because of the time and space requirements for many of our problems, this approach was not available to us. In this respect the permanent library proved to be very valuable. First, it allows us to proceed with the computations in stages, and second, to restart a calculation in case of an unforeseen error provided that intermediate results are kept. Unexpected errors include exceeding the time limit or trying to keep more than 32,767 terms. As the value of the threshold determines how many terms will be kept, it may take some experimenting so that a meaningful number of terms are retained for a series. From a practical standpoint a series with thousands of terms is probably not very useful, and if higher accuracy can be achieved only by computing even more terms, this may be an indication that a different approach may be called for proach may be called for.

3. RESEARCH PROJECTS FOR WHICH POLYPAK HAS BEEN USED

POLYPAK has been used in the following research projects:

1. A discussion of entrainment domains for forced second-order differential equations [6]
2. Formal computation of the limit cycle in van der Pol's equation [3]
3. Computation of a formal solution to Hill's lunar problem with a new way of solving Hill's differential equation [8]
4. Computation of the solution to the main problem of lunar theory [4,9]
5. Normalization of the hamiltonian of the restricted problem of three bodies near the lagrangian equilibrium point L_4 [7]

The most extensive of these projects is the fourth. The first stage of the project is completed and we now work on the perturbations of the motion of moon by the planets.

It was G. Hill who learned from the failure of Delaunay's work. He realized that a different method for computing the motion of the moon was required. Instead of treating the sun as a perturbation of the two-body problem of earth and moon, Hill recognized that the effect of the sun on the moon was too large and should not be neglected in the initial approximation. By including the dominant part of the sun's gravitational force in the differential equation for the initial approximation, he arrived at and was able to solve what is now known as Hill's lunar problem.

It was E. Brown around the turn of the century who carried Hill's ideas to their conclusion by computing the motion of the moon as a perturbation to Hill's lunar problem. Brown gives the answer for the longitude, latitude, and sine parallax in form of multiple Fourier series, although he did the computations with complex coordinates. Today the moon is observed with great precision and Brown's work has fallen into disfavor because of its limited accuracy compared to numerical integration. It has been our research project to redo Brown's calculation by machine with an accuracy that is comparable to the one of the observations. A detailed account of our work appears in Ref. 4.

ACKNOWLEDGMENTS

This research was supported by NSF Grant MCS-8301919. The author would like to acknowledge the inspiration and help of A. Deprit in the development of POLYPAK.

NOTE

Copies of POLYPAK are available from the author upon request. We have used it under the MVS operating system of IBM on the Amdahl 470-V7 of the University of Cincinnati. A knowledge of PL/I is required when POLYPAK is used. The program of a user has to be written in this language. It will be called as the subroutine PROBLEM by POLYPAK.

REFERENCES

1. C. Delaunay, Théorie du mouvement de la lune, Mem. Acad. Sci. *28* (1860); *29* (1867).
2. A. Deprit, J. Henrard, and A. Rom, Analytical lunar ephemeris, Astronom. and Astrophys. *10* (1971), 257.
3. A. Deprit and D. Schmidt, Exact coefficients of the limit cycle in van der Pol's equation, J. Res. Nat. Bur. Standards *84* (1979), 293.
4. M. Gutzwiller and D. Schmidt, *The Motion of the Moon as Computed by the Method of Hill, Brown, and Eckert*, Astronom. Papers, U.S. Naval Observatory, Vol. 23, Part 1 (1986).
5. D. Knuth, *The Art of Computer Programming*, Vol. 3, Addison-Wesley, Reading, Mass., 1973, p. 51.
6. K. Meyer and D. Schmidt, Entrainment domains, Funkcial. Ekvac. *20* (1977), 171.

7. K. Meyer and D. Schmidt, The stability of the Lagrange triangular point and a theorem of Arnold, J. Differential Equations *62* (1986), 222.

8. D. Schmidt, Literal solution for Hill's lunar problem, Celestial Mech. *19* (1979), 279.

9. D. Schmidt, The main problem of lunar theory solved by the method of Brown, Moon and Planets *23* (1980), 135.

Computer Algebra and Definite Integrals

RICHARD ASKEY   University of Wisconsin-Madison, Madison, Wisconsin

It seems to be expected of every pilgrim up the slopes of the mathematical Parnassus that he will at some point or other of his journey sit down and invent a definite integral or two towards the increase of the common stock. Sylvester [21]

Times have changed in the last 125 years and very few mathematicians feel they have not done their duty if they have not discovered (not invented, no matter what Sylvester thought) a new definite integral. In fact, until a few years ago most mathematicians would have been surprised if one claimed that great mathematics might be done by evaluating a definite integral. One of my colleagues even said that he did not think a Ph.D. should be granted if the thesis was just the evaluation of some definite integrals. In the last few years there has been serious work done on the evaluation of some very important multidimensional beta integrals. There are now more conjectures than results and probably much more remains to be discovered that we can conjecture at present.

The deepest result that has been proved is Selberg's beta integral:

$$\int_0^1 \cdots \int_0^1 \prod_{1 \leq i < j \leq n} |t_i - t_j|^{2z} \prod_{i=1}^n t_i^{x-1}(1 - t_i)^{y-1} \, dt$$
$$= \prod_{j=1}^n \frac{\Gamma(x + (j - 1)z)\Gamma(y + (j - 1)z)\Gamma(jz + 1)}{\Gamma(x + y + (n + j - 2)z)\Gamma(z + 1)} \tag{1}$$

See Ref. 20 and an earlier paper (Ref. 19) for a statement of an equivalent integral using the beta integral on $[0,\infty)$ as the basic integral. This result was almost completely lost for over three decades.

The next two integrals were discovered as conjectures by Dyson [6] and Mehta and Dyson [14]. Dyson's conjecture was discovered in the following way. He needed the value of the integral

$$\int_0^{2\pi} \cdots \int_0^{2\pi} \prod_{1 \leq j < k \leq n} |e^{i\theta_j} - e^{i\theta_k}|^{2z} \, d\theta_1 \cdots d\theta_n$$

Using the orthogonality of $e^{in\theta}$ he showed that the evaluation of this integral when $z = m$ is an integer is equivalent to finding the constant term in

$$\prod_{1 \leq j < k \leq n} \left(1 - \frac{x_j}{x_k}\right)^m \left(1 - \frac{x_k}{x_j}\right)^m$$

When $n = 2$ all the terms in this expansion can be found using the binomial theorem. When $n = 3$ the constant term can be found by using Dixon's sum

$$(-1)^m \sum_{j=0}^{2m} (-1)^j \binom{2m}{j}^3 = \frac{(3m)!}{m!\, m!\, m!} \tag{2}$$

Dyson had read Bailey's book [4] and knew that Dixon extended (2) by adding two more parameters. He also knew that more parameters often made it easier to see what was happening and to prove the resulting conjectures. He was able to pick out one way of adding the extra parameters and conjectured that

$$\text{C.T.} \prod_{1 \leq j < k \leq n} \left(1 - \frac{x_j}{x_k}\right)^{a_j} \left(1 - \frac{x_k}{x_j}\right)^{a_k} = \frac{(a_1 + \cdots + a_n)!}{a_1! \cdots a_n!} \tag{3}$$

where

$$\text{C.T.} \ f(x_1, \ldots, x_n)$$

is the constant term in the Laurent series of $f(x_1, \ldots, x_n)$.

This conjecture was proved by Gunson [11] and Wilson [22]. Good [9] abstracted the essence of Wilson's proof to obtain a very elegant proof. Andrews [2] found an integral similar to Selberg's that is equivalent to (3).

The Mehta-Dyson conjecture [14] was

$$\frac{1}{(2\pi)^{n/2}} \int_{-\infty}^{\infty} \cdots \int_{-\infty}^{\infty} \prod_{1 \le i < j \le n} |t_i - t_j|^{2z} \prod_{i=1}^{n} e^{-t_i^2/2} dt_i$$

$$= \prod_{j=1}^{n} \frac{\Gamma(jz+1)}{\Gamma(z+1)} \qquad (4)$$

After Selberg's integral was recalled it was obvious how to obtain (4) from Selberg's integral (1). This is still the only known proof of (4). It would be useful to have a direct one.

The next conjecture was formulated by Andrews [1]. He conjectured that

$$\text{C.T.} \prod_{1 \le j < k \le n} \left(\frac{x_j}{x_k}; q\right)_{a_j} \left(\frac{qx_k}{x_j}; q\right)_{a_k} = \frac{(q;q)_{a_1+\cdots+a_n}}{(q;q)_{a_1} \cdots (q;q)_{a_n}} \qquad (5)$$

where

$$(a;q)_n = (1-a)(1-aq)\cdots(1-aq^{n-1}) \qquad n = 1, 2, \ldots, \quad (a;q)_0 = 1$$

All of these conjectures and results are attractive, but they seemed to be isolated facts. Macdonald discovered a natural setting using root systems and affine root systems. The Dyson-Andrews conjecture (5) comes from the affine root system A_{n-1}. Macdonald had a very attractive conjecture about the constant term of a Laurent polynomial attached to each of the root systems and many of the affine root systems. It had only one degree of freedom as well as the q until Selberg's integral was brought to light and sent to Macdonald. Selberg's integral not only allowed him to prove his conjectures for the root systems B_n, C_n, D_n, and BC_n, but it is equivalent to an extended Macdonald conjecture for BC_n with the roots separated by different parameters, depending on their lengths.

Macdonald's conjecture with one degree of freedom was circulated in a handwritten draft, and it was clearly important to add as many more parameters as possible. The first conjectured extension was found by Morris for G_2. He used his Apple computer to find the constant term with the roots separated by root lengths, and after finding a half dozen or so examples, he was able to formulate the following conjecture:

CONJECTURE 1 (Macdonald a = b, Morris in the general case)

$$\text{C.T. } (1-x)^a(1-x^{-1})^a(1-y)^a(1-y^{-1})^a(1-xy)^a(1-x^{-1}y^{-1})^a$$

$$\left(1-\frac{y}{x}\right)^b\left(1-\frac{x}{y}\right)^b(1-x^2y)^b(1-x^{-2}y^{-1})^b(1-xy^2)^b(1-x^{-1}y^{-2})^b$$

$$= \frac{(3a+3b)!\,(3b)!\,(2a)!\,(2b)!}{(2a+3b)!\,(a+2b)!\,(a+b)!\,a!\,b!\,b!} = G(a,b) \tag{6}$$

This is equivalent to the integral

CONJECTURE 1'

$$\frac{2^{3a+3b-2}}{\pi^2}\int_0^{2\pi}\int_0^{2\pi}(1-\cos\theta)^a(1-\cos\varphi)^a[1-\cos(\theta+\varphi)]^a$$

$$\times [1-\cos(\theta-\varphi)]^b[1-\cos(2\theta+\varphi)]^b[1-\cos(\theta+2\varphi)]^b\,d\theta\,d\varphi$$

$$= G(a,b) \tag{7}$$

Computer algebra was a necessary aid for the next step. When trying to find a q-extension of the Selberg-Macdonald constant term identity for BC_n, it was necessary to use a larger computer and one that had a good symbolic manipulation package. MACSYMA and the MIT computer were used. After finding the constant term for a number of cases for BC_2 and a couple for BC_3, Morris made the following conjecture:

CONJECTURE 2 (Macdonald a = b = c, Morris in the general case)

$$\text{C.T. }\prod_{i=1}^{n}(x_i;q)_a(qx_i^{-1};q)_a(qx_i^2;q^2)_b(qx_i^{-2};q^2)_b$$

$$\times \prod_{1\leq i<j\leq n}(x_ix_j;q)_c(qx_i^{-1}x_j^{-1};q)_c(x_ix_j^{-1};q)_c(qx_jx_i^{-1};q)_c$$

$$= \prod_{j=0}^{n-1}\frac{(q;q)_{2a+2b+2cj}\,(q;q)_{2b+2cj}\,(q;q)_{c(j+1)}}{(q;q)_{a+2b+(n+j-1)c}\,(q;q)_{2b+cj}\,(q;q)_{a+cj}\,(q;q)_c}$$

$$\times \frac{(q^2;q^2)_{2b+cj}\,(q^2;q^2)_{a+cj}}{(q^2;q^2)_{a+b+cj}\,(q^2;q^2)_{b+cj}} \tag{8}$$

The q was very useful in formulating this conjecture. The computer factored the rational function, and the q kept factors from being canceled, as many would have been if q = 1 had been used.

The next case to be considered was F_4 with two degrees of freedom. This problem was too large for the MIT computer using MACSYMA. However,

Morris had supplied Macdonald with enough data and Macdonald was able to reformulate the BC_n conjecture using weights of the roots. His formulation made sense for all the root systems and affine root systems. In particular, there are two affine versions of G_2. One of these has a constant term conjecture which uses products on base q for the short roots and base q^3 for the long roots. Morris worked out the details to obtain the following conjecture.

CONJECTURE 3 (Macdonald-Morris)

$$\text{C.T.} \quad (x;q)_a \left(\frac{q}{x};q\right)_a (y;q)_a \left(\frac{q}{y};q\right)_a (xy;q)_a \left(\frac{q}{xy};q\right)_a \left(\frac{y}{x};q^3\right)_b \left(\frac{q^3 x}{y};q^3\right)_b$$

$$\times \ (x^2 y; q^3)_b \left(\frac{q^3}{x^2 y}; q^3\right)_b (xy^2; q^3)_b \left(\frac{q^3}{xy^2}; q^3\right)_b$$

$$= \frac{(q;q)_{3a+3b} (q;q)_{3b} (q;q)_{2a} (q^3;q^3)_{a+3b} (q^3;q^3)_{2b} (q^3;q^3)_a}{(q;q)_{2a+3b} (q;q)_{a+3b} (q;q)_a^2 (q^3;q^3)_{a+2b} (q^3;q^3)_{a+b} (q^3;q^3)_b^2} \tag{9}$$

Morris looked at A_2 again and found two other ways to label the roots so that the resulting constant term could be evaluated from Dixon's well-poised $_3F_2$ sum [the extension of (2) with two more degrees of freedom]. One of these may be isolated, but Morris showed that the other was one instance of another infinite family of constant term identities, or equivalently of the evaluation of integrals. The integrals are Cauchy beta-type integrals which Selberg had evaluated but never published. Again MACSYMA was used by Morris to discover these conjectures, and only after finding and proving them did he become aware of Selberg's work. MACSYMA also gave enough data so that Morris could formulate a q-conjecture of this type. Macdonald's paper has appeared [13], and Morris's Ph.D. thesis [15] will eventually appear. Copies are available from the present author.

A number of us have spent more time on the Dyson-Andrews conjecture than we like to admit. Kadell proved it for n = 4 [12] and formulated a number of related conjectures. The Dyson-Andrews conjecture was proved by Zeilberger and Bressoud [24], and the related conjectures of Kadell and other results were proved by Bressoud and Goulden [5]. These two papers are very interesting, since the proofs are purely combinatorial. They count tournaments. See Zeilberger [23] for the first proof of this type.

These results and conjectures seem to be just the start. I gave a number of conjectured extensions of Selberg's integral as series [3].

One is

$$\sum_{x_1,\ldots,x_n=0}^{N} \prod_{1\leq i<j\leq n} (1 - k + x_j - x_i)_{2k} \prod_{i=1}^{n} \frac{(\alpha)_{x_i} (\beta)_{N-x_i}}{x_i!(N-x_i)!}$$

$$= \prod_{j=0}^{n-1} \frac{(\alpha)_{jk}(\beta)_{jk}(\alpha+\beta)_{N+jk}(jk)!}{(\alpha+\beta)_{(n+j-1)k}(1)_{n-jk}k!} \qquad (10)$$

with the shifted factorial $(a)_n$ defined by

$$(a)_n = \frac{\Gamma(n+a)}{\Gamma(a)}$$

This paper includes conjectured q-extensions of Selberg's integral of a different type than the Macdonald conjectures. At the end of Ref. 3, I mentioned a number of places where Selberg-type integrals have been used, and raised the question of an extension using Barnes-type integrals. Recall Barnes's beta integral

$$\frac{1}{2\pi} \int_{-\infty}^{\infty} \Gamma(a+it)\Gamma(b+it)\Gamma(c-it)\Gamma(d-it) \, dt$$

$$= \frac{\Gamma(a+c)\Gamma(a+d)\Gamma(b+c)\Gamma(b+d)}{\Gamma(a+b+c+d)} \qquad (11)$$

when $\mathrm{Re}(a,b,c,d) > 0$. Rahman gave a conjectured extension of Selberg's integral somewhat similar to the integral that comes from the Macdonald-Morris conjecture for BC_n, proved it in the two-dimensional case, and showed that it imples the existence of a multidimensional beta integral using a special case of the Barnes beta integral (see Ref. 16).

Jacobi introduced a sum which is now called a Jacobi sum, and evaluated it as a quotient of Gauss sums. Gauss sums are character sum analogs of the gamma function and Jacobi sums are analogous to the beta integral. Selberg in unpublished work evaluated a two-dimensional character sum that is analogous to his integral. Evans [7] published a proof of this, found a variant and proved it in two dimensions, and formulated an n-dimensional conjecture. He gave other conjectures [8], one of which was proved by Greene [10].

There seem to be many more extensions, and as the one-dimensional beta integrals become more complicated, the calculations to find possible n-dimensional versions using the one-dimensional beta-type integral become so large that a computer with the ability to handle algebra is essential. However, there is a need to make these systems less greedy of machine space.

The example of F_4 mentioned above is a case in point. Data could probably have been obtained by writing a system designed specifically for this problem, but the standard systems use too much space to be useful for problems this large.

Some recent applications of these integrals and further integrals were found by Regev (see Refs. 17 and 18). These applications are very surprising, for who would have thought that multidimensional normal integrals would arise from rings with polynomial identities. My experience with identities is that nice ones exist only when there is a good reason for them. If this reason has not yet been found, there is still work to be done. Root systems help explain the existence of Selberg's integral, and affine root systems seem to be responsible for the Macdonald-Morris conjectures. However, there either is no space underlying the other conjectures, or we have not yet found the appropriate ones.

ACKNOWLEDGMENT

This work was supported in part by National Science Foundation Grant MCS-8101568.

REFERENCES

1. G. E. Andrews, Problems and prospects for basic hypergeometric functions, in *Theory and Application of Special Functions*, ed. R. Askey, Academic Press, New York, 1975, pp. 191-224.
2. G. E. Andrews, Notes on the Dyson conjecture, SIAM J. Math. Anal. *11* (1980), 787-792.
3. R. Askey, Some basic hypergeometric extensions of integrals of Selberg and Andrews, SIAM J. Math. Anal. *11* (1980), 938-951.
4. W. N. Bailey, *Generalized Hypergeometric Series*, Cambridge University Press, Cambridge, 1935 (Reprinted: Hafner Press, New York, 1964).
5. D. M. Bressoud and I. P. Goulden, Constant term identities extending the q-Dyson theorem, Trans. Amer. Math. Soc. *291* (1985), 203-228.
6. F. J. Dyson, Statistical theory of the energy levels of complex systems, I, J. Math. Phys. *3* (1962), 140-156.
7. R. Evans, Identities for products of Gauss sums over finite fields, Enseign. Math. *27* (1981), 297-309.
8. R. Evans, Character sum analogues of constant term identities for root systems, Israel J. Math. *46* (1983), 189-196.
9. I. J. Good, Short proof of a conjecture of Dyson, J. Math. Phys. *11* (1970), 1884.

10. J. R. Greene, Character sum analogues for hypergeometric and generalized hypergeometric functions over finite fields, Ph.D. thesis, University of Minnesota, 1984.

11. J. Gunson, Proof of a conjecture by Dyson in the statistical theory of energy levels, J. Math. Phys. *3* (1962), 752-753.

12. K. W. Kadell, Andrews' q-Dyson conjecture: n = 4, Trans. Amer. Math. Soc. *290* (1985), 127-144.

13. I. G. Macdonald, Some conjectures for root systems, SIAM J. Math. Anal. *13* (1982), 988-1007.

14. M. L. Mehta and F. J. Dyson, Statistical theory of the energy levels of complex systems, V, J. Math. Phys. *4* (1963), 713-719.

15. W. G. Morris, II, Constant term identities for finite and affine root systems: Conjectures and theorems, Ph.D. thesis, University of Wisconsin, Madison, 1982.

16. M. Rahman, Another conjectured q-Selberg integral, SIAM J. Math. Anal. *17* (1986), 1267-1279.

17. A. Regev, Asymptotic values for degrees associated with strips of Young diagrams, Adv. in Math. *41* (1981), 115-136.

18. A. Regev, Combinatorial sums, identities and trace identities of the 2 × 2 matrices, Adv. in Math. *46* (1982), 230-240.

19. A. Selberg, Uber einen Satz von A. Gelfand, Arch. Math. Naturvidensk. *44* (1941), 159-170.

20. A. Selberg, Bemerkninger om et Multipelt Integral, Nordisk Mat. Tidskr. *26* (1944), 71-78.

21. J. J. Sylvester, Notes to the meditation on Poncelet's theorem, Philos. Math. XX (1860), 525-533. Also in *Collected Mathematical Papers*, Vol. II, Cambridge University Press, Cambridge, 1908 (Reprinted: Chelsea, New York, 1973, pp. 208-214).

22. K. Wilson, Proof of a conjecture by Dyson, J. Math. Phys. *3* (1962), 1040-1043.

23. D. Zeilberger, A combinatorial proof of Dyson's conjecture, Discrete Math. *41* (1982), 317-321.

24. D. Zeilberger and D. M. Bressoud, A proof of Andrews' q-Dyson conjecture, Discrete Math. *54* (1985), 201-224.

Algebraic Computations and Structures

JAMES H. DAVENPORT University of Bath, Bath, England

This session is devoted to a discussion of those computations that computer algebra can perform, and how they are performed in practical systems. We will discuss some of the common items represented in computer algebra programs, as well as some of the general issues such as canonical representations and sparsity. We will not go into great detail on the algorithms, and we will largely be concerned with questions like "Can X be computed?" rather than "How do we compute X efficiently?"

DATA TYPES OF COMPUTER ALGEBRA

Numbers

There are few theoretical problems in computing with integers [though many in computing with integers in the most efficient way; Knuth (1980) has much to say on the subject]. On the practical side, there is the point that whereas the mathematician believes in the existence of infinitely many integers, the computer designer implements only finitely many (typically, -2^n to $2^n - 1$, where n is 31 for the IBM System/370 range). Hence we cannot directly use the computer's integers to represent the mathematical integers, but must be prepared to write programs to handle arbitrarily large integers represented by means of several computer integers. In fact, there are two major ways of performing such a representation: we can choose either a *radix* notation (a generalization of conventional decimal notation), in which n is represented as $\sum_{i=0}^{k} n_i B^i$, where $-B/2 < n_i \leq B/2$ and B is representable in a single computer word, or a *modular* notation, in which n is represented by its value modulo a sufficient number of large (but representable in one computer word) primes. There are interesting trade-offs between these two representations: the modular form is much faster for arithmetic operations (addition, subtraction, multiplication) since the individual radices are completely independent, but much slower for comparison, or for deciding if one number divides another. Hence the choice of representation influences the algorithms that will be chosen.

Similarly, rational numbers present few conceptual difficulties, although great ingenuity has been expended to minimize the actual time taken on a computer to manipulate them. Rational numbers do, though, bring out one important idea—that of *canonical representations*. We say that a domain has canonical representations if two elements are equal iff they have the same representation. Canonical representations are one of the most important techniques in computer algebra*, and great effort has been, and is still being, expended to discover canonical representations for certain classes of expressions. To achieve canonical representation for the rational numbers, we need to ensure that the rational numbers are in lowest terms, and also that the denominator is positive (although other choices, such as numerator positive, are possible), for else we would have 2/1 and -2/-1 equal but having different representations.

Algebraic numbers present great theoretical and practical difficulties, but since these problems are common to all algebraic extensions, we defer our discussion of them until we mention general algebraic extensions. The numbers modulo p are very important in computer algebra—not only are they important mathematical objects in their own right, but, as a finite model of the integers, many of the key algorithms of computer algebra rely heavily on them.

Polynomials

Univariate polynomials do not pose any great theoretical problems when it comes to addition or multiplication. Here too, though, there is substantial interest in the question of how to multiply them efficiently, and

---

*Indeed, Loos (1974) regards their existence as essential for a formal theory of computer algebra.

again we refer the reader to Knuth (1980). There is, however, one very important issue that we will mention here, for it crops up throughout computer algebra, and that is *sparsity*. A polynomial (of degree n) is said to be sparse if most of the coefficients between 0 and n are zero (this is deliberately not a rigorous definition—"Sparsity is in the eye of the beholder"). In informal terms we can note that sparsity is preserved by addition and subtraction, and somewhat preserved by multiplication, while division can totally destroy sparsity [consider $(x^n - 1)/(x - 1)$]. An interesting question, which is even hard to formulate precisely, is whether the process of taking greatest common divisors preserves sparsity [see Zippel (1979,1981)]. Factoring $x^n \pm 1$ clearly does not preserve sparsity, but it is a remarkable fact that factoring other binomials does. See Davenport (1983) for this result and further references, and Bremner (1981) for a remarkable trinomial factorization.

A representation scheme for polynomials is said to be sparse[*] if it takes advantage of sparsity in the polynomials it is storing, so that a sparse polynomial occupies less storage space than a dense one of the same degree. A typical dense representation would represent a polynomial of degree n by a list of n + 1 coefficients, while a sparse representation would represent it by a list of ordered pairs (nonzero coefficient, corresponding exponent), so that $x^{1000} + 1$ would be represented as $(1,0,0,\ldots,0,1)$ for 1001 terms in the dense representation, but as $((1,1000)(1,0))$ in the sparse representation. If we consider the "naive" algorithms for multiplying polynomials (and although many asymptotically faster algorithms exist, I know of no system that actually uses them, largely because these algorithms do not become faster until the inputs are far[†] larger than anything met in practice, and would not fit in present computers anyway) as applied to two polynomials of degree n with n' nonzero coefficients, the dense algorithms will require $\Theta((n + 1)^2)$ operations, and the sparse one will require $\Theta(n'^2 \log n')$ operations[‡]. Most computer algebra systems in existence today use sparse representations for polynomials, not only because

---

[*]Note that this language is perverse inasmuch as, for a sparse polynomial, a sparse representation packs the information (this can be made precise, but I do not intend to) more densely than a dense representation.

[†]But see Probst and Alagar (1982) for a further discussion of this issue and some examples.

[‡]The extra logarithmic term is required to sort the output into descending order of degree.

that allows the system to compute $(x^{100,000} - 1)(x^{100,000} + 1)$, which, although trivial, makes one feel fairly foolish if one cannot do, but also for the more cogent reason that practically all multivariate polynomials which actually appear in computer algebra problems are sparse (or else they would be too big to store—even a polynomial of degree 5 in five variables has 7776 terms*).

We can compute the quotient of two polynomials over a field, assuming that one exactly divides the other, and we can produce a quotient and remainder if it does not. If we are not operating over a field, the definition of division by nonmonic polynomials is rather vague—in Z[x], what is

$$\frac{x^2 - 1}{2x - 2}?$$

Questions of greatest common divisors, factoring, and so on, raise more general issues of constructability and we will return to these later.

A multivariate polynomial can always be considered as a univariate polynomial whose coefficients are multivariate polynomials in one fewer variables; and hence we can reduce problems in multivariate polynomials to problems in univariate polynomials. When it comes to division, though, the order in which we chose to split off the univariate extensions is all important:

$$\frac{x + 2y}{x + y}$$

is 1 remainder y in Z[y][x], but 2 remainder -x in Z[x][y]. A related question is that of computation with ideals in multivariate polynomial rings. Here the key process seems to be the computation of Gröbner bases (Buchberger, 1979, 1983), for which a multivariate representation seems essential.

Rational functions give rise to few problems beyond those of polynomials and rational numbers, since they are formed from the polynomials the same way as the rational numbers are formed from the integers—by means of the instruction "take quotient field." To keep the rational functions in canonical form, we need to take greatest common divisors of the polynomials involved, and this problem is discussed below. In addition, we need some condition similar to that used for rational numbers, which

---

*But see Stoutemyer (1984) for a slightly different point of view and a comparison of several representations.

guarantees a unique choice among elements related by units. In the case of rational functions over the integers, the choice "leading coefficient of the denominator positive" is usual, although others are possible.

Algebraic Functions

Algebraic functions (in which category we can include algebraic numbers) can cause all sorts of problems, ranging from the purely mathematical to the purely programming. One of the major ones is the difficulty of ensuring canonical representations; both $\sqrt{6}$ and $\sqrt{2}\sqrt{3}$ represent the same value. This, of course, is because $x^2 - 6$ factorizes in $Z(\sqrt{2},\sqrt{3})$.

Even if we have only one algebraic expression, though, the problem of canonical representations is not trivial—consider the fact that

$$\sqrt{2} - 1 = \frac{1}{\sqrt{2} + 1}$$

even though $\sqrt{2} + 1$ is monic when regarded as a polynomial in $\sqrt{2}$. There are a variety of solutions to this problem. One [discussed at greater length by Davenport (1981)] is to insist that the denominator be free of all algebraic expressions (by multiplying top and bottom by the conjugate of the denominator) and then that it be monic and that there should be no nonalgebraic expression by which we can cancel*.

Matrices

Matrices occur very often in the statement of computer algebra problems, largely because linear equations are generally stated in terms of them. The "obvious" representation of a matrix as a two-dimensional array of coefficients (often in itself represented as a vector of vectors) is indeed efficient for arbitrary matrices, but then many "interesting" matrices are of special forms, such as symmetric or tridiagonal. Whether we use the "obvious" representation or one suited to the special nature of our matrix, the basic operations of addition, multiplication, and so on, are straightforward to implement.

Here, too, though, the issue of sparsity raises its head. Many of the matrices one encounters are sparse (i.e., most elements are zero).

---

*More formally, if the denominator d lies in K and the numerator n in an algebraic extension $K[x_1,\ldots,x_k]$ of K, we require that $\gcd(d,n') = 1$ (computed in K), where n' is the content of n in the variables $x_1, \ldots, x_k$.

There are many good reasons for this, one being that many matrices are derived from graphs, and then elements are nonzero only if the row and column are connected directly in the graph. A common and suitable representation for sparse matrices is as a vector of rows, where each row is a list of pairs (column position, element belonging to row and column) and where only nonzero elements are stored explicitly.

One of the main reasons for working with matrices is to solve linear equations, whether as an end in themselves or as part of a broader algorithm (e.g., Berlekamp [1967], where the factorization of polynomials over finite fields is achieved by constructing and solving suitable sets of linear equations). Here the algorithms split into two completely different classes, depending on whether or not the system is sparse (as mentioned above, this is not a rigorous distinction, because the algorithms will work in either case; they are just more efficient in the case for which they were designed).

In the dense case, most algorithms for solving linear equations (or finding inverses or determinants of matrices, for the tasks are closely linked), for matrices with no special structure, are based on gaussian elimination, as they are in numerical analysis. The main concern of numerical analysts in this process is with the numerical stability of the process (i.e., the accuracy of the intermediate results). In computer algebra, of course, the results are always accurate, and the major concern is over the size of the intermediate answers, or *intermediate expression swell*, as it is often termed. A good example of this is the matrix

$$\begin{bmatrix} x-1 & y-1 & z-1 & 1 \\ y-2 & z-2 & x-2 & 1 \\ z-3 & x-3 & y-3 & 1 \\ x+y+z-6 & x+y+z-6 & x+y+z-6 & 3 \end{bmatrix}$$

whose determinant is zero, but which gives rise to such intermediate expressions as

$$\frac{xz - 3x - y^2 + 5y - 2z}{-x^3 + 6x^2 + 3xyz - 6xy - 6xz - y^3 + 6y^2 - 6yz - z^3 + 6z^2}$$

Not only are these expressions large (compared with the input and with the final answer), but they are also the quotients of large polynomials, and hence require the computation of greatest common divisors to keep them in a canonical form (and if we did not reduce by the greatest common divisors, they would probably be even larger). Hence there is substantial interest in methods that reduce both of these disadvantages, and a useful method [due to Bareiss (1968) and implemented algebraically by Lipson (1969)] combines a two-step technique for reducing the size of the expression with the "fraction-free" idea, which says, essentially, that all the divisions except those that we know to be exact should not be performed, merely remembered and carried out only at the end (if by then they are not exact). There is an interesting variant of this idea, due to Sasaki and Murao (1982), which does gaussian elimination with division but performs the division by power-series expansion in some new variables followed by truncation. Unpublished measurements by the author indicate that this method is often faster than gaussian elimination, although often slower than expansion by nested minors.

In the sparse case, on the other hand, gaussian elimination is disastrously inefficient*, and the methods of choice are based on Cramer's rule (i.e., expansion by minors). There is also great scope for improving on a naive implementation of this expansion, and indeed much work has been done on this [see Smit (1979) for a survey of the state of the art].

2. CONSTRUCTIVE ALGEBRA?

Computer algebra is different from "pure" algebra inasmuch as one generally wants a result rather then merely the knowledge that a result exists. This apparently petty distinction colors the entire field and could almost be said to lead to a different field—constructive algebra. To illustrate this, let us compare the definition of "group" in algebra with a possible definition of "constructive group."

---

*Largely because the inverse of a sparse matrix can be dense, for example,

$$\begin{bmatrix} 3 & 1 & 0 & \cdots & 1 \\ 1 & 1 & 0 & & 0 \\ 0 & 0 & 1 & & \cdot \\ \cdot & \cdot & & & 0 \\ 1 & 0 & 0 & \cdots & 1 \end{bmatrix}$$

and hence should not be calculated (as gaussian elimination techniques for determinants and the solution of linear equations do) unless it is really needed.

DEFINITION 2.1 A group G is a system of elements that is closed under a single-valued binary operation which is associative, and relative to which G contains an element satisfying the identity law, and with each element another element (called its inverse) satisfying the inverse law (Birkoff and MacLane, 1953, p. 126).

DEFINITION 2.2 A set of elements G is said to form a constructive group if there are three operations*: identity, inverse, and times, defined on that set such that:

1. $\forall\ x \in G$ times(x,identity()) = x = times(identity(),x)
2. $\forall\ x \in G$ times(x,inverse(x)) = identity() = times(inverse(x),x)
3. $\forall\ x,y,z \in G$ times(x,times(yz)) = times(times(x,y),z)

Here the difference does not appear too substantial, and it might appear that "constructive algebra" was merely a new terminology. The difference really shows up, though, in the problem described below.

GCD and Factorization

Define a *gcd domain* R to be an integral domain such that every chain "d_1 is divisible by d_2...is divisible by d_n..." consists of associates (i.e., the quotients are units) from some point on (the *divisor chain condition*), and such that every pair of elements has a greatest common divisor (i.e., $\forall\ a,b \in R\ \exists\ c \in R$ such that $c|a, c|b$, and if $d|a$ and $d|b$, then $d|c$). Define a *unique factorization domain* to be an integral domain such that every element has a unique factorization into primes (i.e., every element x of R can be expressed as $p_1 p_2 \cdots p_n$ where the p_i are irreducible, in essentially only one way). Then (Jacobsen, 1974, Th. 2.22) the two concepts are equivalent.

As a constructive analog, define a constructive gcd domain to be a ring satisfying the divisor chain condition complete with an operator gcd satisfying the appropriate rules [gcd(x,y)|x, gcd(x,y)|y, and d|x, d|y implies d|gcd(x,y)]. Similarly, define a constructive unique factorization domain to be a ring complete with an operator factor taking an element of the ring into a list of irreducible elements of the ring, satisfying all the axioms of a (nonconstructive) unique factorization domain.

---

*Looking at it from the viewpoint of a computer programmer, we could say that a data type G forms a constructive group iff there are three functions defined on the data type G satisfying the conditions.

Algebraic Computations and Structures

Then indeed a constructive unique factorization domain is a constructive gcd domain, for the gcd of two numbers is the product of their common factors [after allowing for associates, so that, for example (in the integers), if one number had a factor of 2, and the other one of -2, that would count as a common factor]. However, the converse is not easily proved, and is actually false. This is because, in the classical setting, the first step of the proof remarks that every element of R is indeed a product of irreducibles (for if it is not itself irreducible, it is the product of two elements, and by the divisor chain condition, this reduction terminates after a finite number of steps in a product of irreducibles). This step is not constructive, and since it relies on the divisor chain condition, which is in itself nonconstructive, it is hard to see how to make it constructive.

As an illustration of the difference between the greatest common divisor problem and the factorization problem, we can consider the discrepancy between the following:

1. It is possible to take the greatest common divisor of two numbers less than N in time $O(\log^2 N)$ (Knuth, 1980, Sec. 4.5.3).
2. Factorizing a number less than N can take time $O(e^{\sqrt{6 \log N \log \log N}})$ by one of the best known methods (Schnorr, 1982).

GCDs of Polynomials

One particular result (adapted from a well-known theorem in algebra) is:

THEOREM 2.1 If R is a constructive gcd domain, and x is transcendental over R, then R[x] is a constructive gcd domain.

Proof (Based on Euclid's algorithm): The divisor chain condition is easy, since if d_1 is divisible by d_2, the leading coefficient of d_1 is divisible by the leading coefficient of d_2, so we can apply the divisor chain condition in R to the leading coefficients of the d_i.

So let A_0 and B_0 be two elements of R[x]. We must show how to compute $\gcd(A_0, B_0)$. Without loss of generality we can assume that the degree of A_0 is not smaller than that of B_0. Then consider the sequence of pairs (A_0, B_0), (A_1, B_1), ..., where each pair (A_i, B_i) is obtained from the preceding one by letting A_i and B_i be B_{i-1} and

$$\text{lc}(B_{i-1})A_{i-1} - x^{\deg(A_{i-1}) - \deg(B_{i-1})} \text{lc}(A_{i-1})B_{i-1} \tag{1}$$

(where we use lc to stand for the leading coefficient of a polynomial, and deg to stand for its degree) in some order chosen so as to ensure that A_i is not of lower degree than B_i. The sequence of distinct polynomials formed this way is termed a *polynomial remainder sequence* because (to within a factor from R) each term is the remainder of dividing the previous two. Note that this is an effective computation, using only addition, subtraction, and multiplication in R[x]. The total degree of the pair (A_i, B_i) is less than that of the pair (A_{i-1}, B_{i-1}), since the expression (1) is of lower degree than A_{i-1}, as the leading coefficient of A_{i-1} has been canceled. Hence the sequence must terminate with some B_n being zero. Now let A'_n be the primitive part of A_n [i.e., $A_n/\text{content}(A_n)$] (this, too, is effective, since the content is computed as the greatest common divisor of certain elements of R which was assumed to be a constructive gcd domain). Then, since $B_n = 0$, we have that A'_n divides A_{n-1}, for equation (1) can be written in this case as

$$\text{lc}(B_{n-1})A_{n-1} - x^{\deg(A_{n-1})-\deg(B_{n-1})} \text{lc}(A_{n-1})B_{n-1} = B_n = 0$$

and if we let c stand for the content of $A_n = B_{n-1}$, we can write this as

$$c_1 A_{n-1} = x^k c_2 c A'_n$$

for $c_1, c_2 \in R$, k a nonnegative integer. Since A'_n is primitive, this implies that A'_n divides A_{n-1}. But we know that it divides B_{n-1}. Since the elements of the pair (A_{n-2}, B_{n-2}) are linear combinations of A_{n-1} and B_{n-1}, we know that A'_n divides both of them, and by induction we see that it divides all the A_i and B_i. Hence A'_n is a divisor of A_0 and B_0. Now let C be $A'_n \gcd(\text{content}(A_0, B_0))$, again an effectively computable quantity. Since A'_n divides both A_0 and B_0, we can deduce that C does, and in fact it is their greatest common divisor.

For let Q divide both A_0 and B_0, and write Q as qQ', where q is the content of Q. Then q certainly divides C, so the only problem might be Q'. Since A_i and B_i are linear combinations of A_{i-1} and B_{i-1}, we can deduce that if Q' divides A_{i-1} and B_{i-1}, it divides A_i and B_i. Hence inductively Q' divides A_n, and hence (since it is primitive) it divides A'_n, so we deduce that Q divides C, as required.

This proof has been given at some length to show how a theory of constructive algebra might begin. Although the algorithm given (essentially repeated application of equation (1), followed by fixing up the content of the answer] might appear simple, it is hard to think of any algorithm on which so much has been written. The subject of efficient polynomial remainder sequences has generated a great deal of research [for what I believe to be the latest contribution, see Hearn (1979)], and there is no sign that the subject is closed.* Although space does not permit us to go into the details of these variants,[†] and their advantages and disadvantages, we can explain the basic scope for improvement. We are continually multiplying by elements of R, so that even if A_0 and B_0 were co-prime (and since we have to calculate their contents anyway, we might as well make them co-prime), it is likely that the later stages of the algorithm would involve polynomials with large contents (and if we are manipulating polynomials in n variables, the contents will be polynomials in n - 1 variables). We could make the A_i and B_i primitive at each step (the so-called *primitive remainder sequence* algorithm), but this involves many gcd calculations in R. Hence one common approach is to analyze the operation of the algorithm carefully in order to deduce some factor which one knows will divide all the coefficients, and to cancel that.

The obvious question to ask next is whether an analogous theorem holds for constructive unique factorization domains. It does not, and the following counterexample is due to Frohlich and Shepherdson (1956). Let us first consider a trivial example to show how factoring can encode many types of information. Define a_n to be 1 if 2n is the sum of two primes, and -1 otherwise. Let L be the field $\mathbb{Q}(\sqrt{a_n} \; \forall n)$. Then L is either $\mathbb{Q}$ or $\mathbb{Q}[i]$, and hence L[x] is a unique factorization domain and a constructive gcd domain. However, it is not a constructive unique factorization domain, since factoring $x^2 + 1$ is equivalent to solving Goldbach's conjecture, since $x^2 + 1$ is irreducible if i is not in L (i.e., every even number is the sum of two primes).

---

*The competition between these methods and methods based on the Chinese remainder theorem (modular methods) and on Hensel's lemma (p-adic methods) probably spurs on many of the researchers. The fact that continual improvements are still being made in such a basic area as the greatest common divisors of polynomials shows that there is much still to be done in computer algebra. Indeed, while this paper was being written, a new contender arrived, a heuristic algorithm based on substituting an integer for the indeterminate (Char et al., 1984).

[†]See Loos (1982) for a recent treatment.

After this example, it is difficult to see how any of the methods currently known for factorizing polynomials over the integers can be generalized to arbitrary R. There are basically two families of such algorithms—those based on Kronecker's method (van der Waerden, 1940, Vol. 1, p. 77) and those based on Berlekamp's algorithm and Hensel's lemma (Moses and Yun, 1973). The former seem to depend inherently on their being only a finite number of units in R (indeed, van der Waerden makes this an explicit condition), and the latter depend heavily on reducing the problem to the integers modulo p.

Let $S = \{f(n)\}$ be a recursively emunerable set of natural numbers that is not recursive.* Let $L = \mathbb{Q}(\sqrt{p_f(1)}, \sqrt{p_f(2)}, \ldots)$, where p_i is the ith prime. Then the polynomial $x^2 - p_k$ is reducible in $L[x]$ if and only if $k \in S$, and there can be no program to decide this.

Differences from "Constructive Mathematics"

There has sprung up the field of constructive mathematics, where a proof is valid only if each step consists purely of constructive operations. We can quote Seidenberg (1974) here, as an illustration of the guiding spirit of constructive mathematics: "In this section we show how to remove every non-finite form of reasoning from our constructions. The constructions themselves are, of course, already in finite terms, but the underlying theory is not." This approach is actually too limiting for computer algebra, which will accept a nonconstructive proof of a constructive algorithm.

It is hard to give a trivial example of such a proof, but here is the sketch of one (taken from Davenport, 1981, Chap. 6). Given a divisor D on an algebraic curve lying over K(u) for some field K and u transcendental over K, we wish to determine the order of D if it is of finite order. There is an algorithm, based on the theory of Gauss-Manin operators, for determining whether or not the divisor is of finite degree, but it tells us nothing about the degree—to that extent it is nonconstructive. But once we know that it is of finite degree, we can search the positive integers for it, since for any particular n we can determine if D is of order n or not. Hence we have converted a nonconstructive proof of finite order into a constructive algorithm for determining the order.

---

*This means that we can write a computer program to emunerate the set S [i.e., as f(1), f(2), ...] but not one which will decide if a given integer belongs to S. Such sets exist [see Frohlich and Shepherdson (1956)].

3. POWER SERIES

So far we have concentrated almost exclusively on those algebraic operations which can be carried out precisely, and on those expressions with a finite (and hence manipulable) representation. But the sciences, and much of applied mathematics, wish to deal with things that cannot be computed precisely (e.g., the solution to a differential equation). The traditional tool for this has been the series solution (normally, though by no means inevitably, the power series). Such series can be extremely tedious to develop; indeed, much of the early development of computer algebra was directed toward mechanizing this process.

The Traditional Approach

The "traditional" approach to the representation of power series has been to truncate at some specified point, and then to regard them essentially as polynomials. Most computer algebra systems provide some facility for doing this, and in many it is sophisticated, with facilities for declaring, say, that ε is first order, while δ is second order, and that the entire expression is to be maintained up to tenth-order terms, say.

The major difficulties with this approach are ones it shares with numerical analysis in general (and indeed the two concepts are very similar, with "number of terms" corresponding to "number of decimal places"): inflexibility and cancellation. The former can be seen immediately from the fact that all these traditional systems require users to specify the degree to which they wish to work in advance, whereas in practice they are quite likely to look at the answers and then wish for another term, which cannot be computed except by redoing all the previous calculations.

Although extremely irritating in practice, this drawback is far less serious than that of cancellation. As an illustration, suppose that $f(x) = 1 + \sum_{n=2}^{x} a_r x^i$ (i.e., the x term is missing) and consider $\sqrt{f(x) - 1}$. This is then

$$\sqrt{a_2}\, x + \frac{a_3}{2\sqrt{a_2}} x^2 + \frac{4 a_2 a_4 - a_3^2}{8 a_2^{3/2}} x^3 + \cdots$$

Hence if we truncate $f(x)$ at the n term $a_n x^n$, we can only deduce the expansion of $\sqrt{f(x) - 1}$ as far as the x^{n-1} term, for the next term would involve a_{n+1}. Not only will this lessen the accuracy of our result, but we may not even know about it, since our polynomial-handling routines will just report that the x^n term has a coefficient of zero. This means, then, that it is just as possible to get rubbish when computing with power series as it is when computing with any other form of approximation (e.g., floating-point numbers).

Norman's Technique

To overcome both these disadvantages, Norman (1975) introduced a technique of representing a power series by its generating function rather than by any finite list of coefficients, and Norman summarizes the design by saying that his package "gives the impression that it is computing with full rather than truncated power series. Calculations are only performed when an attempt is made to display or use the results, and so no unnecessary work is ever done." This sounds wonderful—how is it achieved?

In a simple case, the generating process is simple, so that if $f(x) = g(x) + h(x)$, we would generate f_i as $g_i + h_i$. Subtraction and multiplication are similarly easy, but what about division? Since this case shows the general method well, we will consider this in some detail. The first remark is that it is sufficient to solve the problem of finding reciprocals, for then division follows from that and multiplication. So suppose that we are given $f(x) = \Sigma_{i=0}^{\infty} f_r x^i$, and wish to compute $g = 1/f$. We shall assume that $f_0 \neq 0$, for else we need a Laurent series, and, while theoretically no more difficult, the exposition gets more complex. Then, since $fg = 1$, we have that $f_0 g_0 = 1$, and that $\Sigma_{i=0}^{n} g_i f_{n-i} = 0$ ($n \geq 0$). So we can deduce that $g_0 = 1/f_0$ and that

$$g_n = \frac{-\Sigma_{i=0}^{n-1} g_i f_{n-i}}{f_0}$$

This is a perfectly effective definition of g_n in terms of f and previously computed* terms of g, so we have the generating function required.

Similar relationships can be found for all functions defined by differential equations, and we can also use this method to deduce power series expansion for "new" functions that we meet. The method can also be used to convert a power series for y in terms of x into one for x in terms of y.

---

*Because of the heavy reliance on previously computed terms, it is in practice necessary to store all computed terms rather than working them out afresh. In this case (assuming that the f_i are known) this reduces the cost of computing g_n from $O(2^n)$ to $O(n)$.

It is worth pointing out that these techniques are not limited to the applied mathematician: one recent approach to computer algebra on algebraic curves (Davenport, 1981) uses Puiseux expansions as the major tool for manipulating functions. Here, of course, many of the "interesting" calculations have to be done at singularities, where a great deal of cancellation arises.

REFERENCES

Bareiss, E. H., Sylvester's identity and multistep integer-preserving Gaussian elimination, Math. Comp. *22* (1968), 565-578.

Berlekamp, E. R., Factoring polynomials over finite fields, Bell System Tech. J. *46* (1967), 1853-1859.

Birkhoff, G., and MacLane, S., *A Survey of Modern Algebra*, Rev. ed., MacMillan, New York, 1953.

Bremner, A., On reducibility of trinomials, Glasgow Math. J. *22* (1981), 155-156.

Buchberger, B., *A Criterion for Detecting Unnecessary Reductions in the Construction of Groebner Bases*, Proc. EUROSAM 79 (Lecture Notes in Computer Science 72), Springer-Verlag, Berlin, 1979.

Buchberger, B., *Groebner Bases: An Algorithmic Method in Polynomial Ideal Theory*, CAMP Publication 83-29.0, Universitat Linz, Austria, November 1983.

Char, B. W., Geddes, K. O., and Gonnet, G. H., *GCDHEU: Heuristic Polynomial GCD Algorithm Based on Integer GCD Computation*, Proc. EUROSAM 84 (Lecture Notes in Computer Science 174), Springer-Verlag, Berlin, 1984.

Davenport, J. H., *On the Integration of Algebraic Functions* (Lecture Notes in Computer Science 102), Springer-Verlag, Berlin, 1981.

Davenport, J. H., *Factorization of Sparse Polynomials*, Proc. EUROCAL 83 (Lecture Notes in Computer Science 162), Springer-Verlag, Berlin, 1983, pp. 214-224.

Frohlich, A., and Shepherdson, J. C., Effective procedures in field theory, Philos. Trans. Roy. Soc. London Ser. A *248* (1955-56), pp. 407-432.

Hearn, A. C., *Non-modular Computation of Polynomial Gcds Using Trial Division*, Proc. EUROSAM 79 (Lecture Notes in Computer Science 72), Springer-Verlag, Berlin, 1979, pp. 227-239.

Jacobsen, N., *Basic Algebra I*, W. H. Freeman, San Francisco, 1974.

Knuth, D. E., *The Art of Computer Programming*, Vol. II, *Seminumerical Algorithms*, 2nd ed., Addison-Wesley, Reading, Mass., 1980.

Lipson, J. D., *Symbolic Methods for the Computer Solution of Linear Equations with Applications to Flow-Graphs*, Proc. 1968 Summer Institute on Symbolic Mathematical Computation, IBM, Yorktown Heights, N.Y., 1969, pp. 233-303.

Loos, R. G. K., Towards a formal implementation of computer algebra (Proc. EUROSAM 74), SIGSAM Bull. *8*(3) (1974).

Loos, R. G. K., *Generalised Polynomial Remainder Sequences*, Symbolic and Algebraic Computation (Supplementum 4), ed. B. Buchberger, G. E. Collins, and R. G. K. Loos, Springer-Verlag, Vienna-New York, 1982, pp. 173-187.

Moses, J., and Yun, D. Y. Y., The EZ GCD algorithm, Proc. ACM *73* (1973), 159-166.

Norman, A. C., Computing with formal power series, ACM Trans. Math. Software *1* (1975), 346-356.

Probst, D., and Alagar, V. S., *An Adaptive Hybrid Algorithm for Multiplying Dense Polynomials*, Proc. EUROCAL 82 (Lecture Notes in Computer Science 144), Springer-Verlag, Berlin, 1982, pp. 16-23.

Sasaki, T., and Murao, H., Efficient Gaussian elimination method for symbolic determinants and linear systems, ACM Trans. Math. Software *8* (1982), 227-289.

Schnorr, C. P., Refined analysis and improvement on some factoring algorithms, J. Algorithms *3* (1982), 101-127.

Seidenberg, A., Constructions in algebra, Trans. Amer. Math. Soc. *197* (1974), 273-313.

Smit, J., *New Recursive Minor Expansion Algorithms, A Presentation in a Comparative Context*, Proc. EUROSAM 79 (Lecture Notes in Computer Science 72), Springer-Verlag, Berlin, 1979, pp. 74-87.

Stoutemyer, D. R., *Which Polynomial Representation Is Best?* Proc. 1984 Macsyma Users' Conference, pp. 221-244.

van der Waerden, B. L., *Modern Algebra*, Julius Springer, Berlin, 1940 (English translation, Frederick Ungar, New York, 1953).

Zippel, R. E., *Probabilistic Algorithms for Sparse Polynomials*, Proc. EUROSAM 79 (Lecture Notes in Computer Science 72), Springer-Verlag, Berlin, 1979, pp. 216-226.

Zippel, R. E., *Newton's Iteration and the Sparse Hensel Algorithm*, Proc. SYMSAC 81, Association for Computing Machinery, New York, pp. 68-72.

Computation of Galois Groups from Polynomials over the Rationals

DAVID J. FORD and JOHN McKAY   Concordia University, Montreal, Quebec, Canada

1. INTRODUCTION

The determination of the Galois group $\text{Gal}_Q f$ from a polynomial f with rational (or integer) coefficients would appear to be a topic of interest to devotees of algebraic and symbolic computation. However, there have been few publications on this problem during the past 20 years.

A glance at van der Waerden [vdW, Sec. 8.10, p. 197] shows that the problem has been solved in the mathematical sense that there is an algorithm available but (and this illustrates a difference between mathematics and computer science) it is not one that leads to a feasible computation since it requires factoring a polynomial of degree n! in n + 1 variables.

It is thought-provoking that few graduate students would know how to approach the question of determining $\text{Gal}_Q f$ for, say, $f(x) = x^6 + 2x^5 + 3x^4 + 4x^3 + 5x^2 + 6x + 7$. Almost all examples are chosen so that the Galois action is obvious and $\text{Gal}_Q f$ is solvable, or $\text{Gal}_Q f$ acts naturally as the symmetric or alternating group and special properties of these groups are exploited.

After writing [EFM] it became apparent that there are a few principles which lead to a powerful uniform approach to finding $\text{Gal}_Q f$. These are:

1. The computation should be exact and constitute a proof of the result.
2. A polynomial determines $\text{Gal}_Q f$, not as a permutation group on the roots

but as a class of permutation groups to within relabeling of the roots. This class is the conjugacy class of $\text{Gal}_Q f$ in the symmetric group.
3. Using the theory of symmetric functions we may find other polynomials whose coefficients are symmetric functions of the roots of f.
4. Factoring the polynomials produced above over the rationals, we show that certain nonsymmetric functions of the roots of the original rational f (which occur as coefficients of factors of the new polynomials) are integer valued.

2. THE METHOD

Suppose that we were to assign the roots according to their approximate numerical values. Then we would need to check whether functions such as $x_1 x_2 x_4 + x_2 x_3 x_5 + x_3 x_4 x_6 + x_4 x_5 x_7 + x_5 x_6 x_1 + x_6 x_7 x_2 + x_7 x_1 x_3$ [which occurs in [EFM] and is associated with the group PSL(3,2) and the Steiner system S(2,3,7)] have integer values when evaluated on the roots of f. It may be apparent that the result is not an integer since the difference between the numerical value and an integer exceeds the known error in the computation, in which case we relabel the roots and try again. If, however, it is clear that the value is very close to an integer, it is difficult to *prove* that it is an integer. One way of doing so is to use Baker's results on approximating linear forms in logarithms of algebraic integers, but this requires the values of the roots to (typically) many hundreds or thousands of decimal digits.

What is needed is to use invariants—not of some permutation group as above, but the class of the Galois group [KM]. Such invariants are the shapes of elements as disjoint products of cycles; these shapes are partitions of the degree of the polynomial (or equivalently, the degree of the group acting as a group of permutations on the roots). Other invariants are the orbit lengths of the group acting on unordered subsets of the roots taken r at a time. Similarly, we may work with ordered subsets. Soicher [S1, S2, SMc] computes these invariants relying on resultant theory for the computations. Stated simply, the shapes of elements are determined by factoring f modulo good primes (i.e., primes not dividing the discriminant of f). The orbit lengths are given by the degrees of the irreducible factors over the rationals of the polynomial, the roots of which are sums of the roots of f taken r at a time. Provided that no two sums of the roots taken r at a time have the same numerical value (and this can always

Computation of Galois Groups

be ensured by making a Tchirnhaus transformation if necessary), the action of f on r-sets is that on r-sums. A group transitive on r-sets is called r-fold set transitive, which is distinct from the more usual notion of r-fold transitivity corresponding to transitivity on ordered r-sets (see [C]); this action is that of $\text{Gal}_Q f$ acting on fixed integer linear combinations of roots taken r at a time (as described in [S2]), again with the proviso that their values be pairwise distinct.

Butler and McKay [BM] tabulate the transitive permutation groups of degree up to 11 giving generators and the distribution of shapes of elements. McKay and Regener [McR] give the orbit lengths for these groups. These invariants, together with the discriminant, do not determine the group in general (even up to isomorphism), but for groups up to degree 11 there are few exceptions, and these have to be treated specially.

Transitive groups for $\text{Gal}_Q f$ correspond to f being irreducible, which assumption is made with some loss of generality since reducible f may require determination of intersections of the splitting fields of factors.

An outline of the flow of the computation is as follows:

I. Is f irreducible over the rationals?
D. Compute the discriminant of f.
C. Factorize f mod good primes r until the stopping criterion is met. $\text{Gal}_Q f$ found?
O. Compute orbit lengths on r-sets. $\text{Gal}_Q f$ found?
A. Use ad hoc methods.

Some comments are in order—it may be preferable to compute D before I. There are several ways of computing the discriminant; one is to compute the resultant of f and f' using a polynomial remainder sequence; another is to compute det VV^t, where $V_{ij} = x_j^{i-1}$, and use Newton's relations to express the power sum entries in terms of the coefficients of f; yet another is to express the discriminant explicitly as a sum of products of the coefficients of f. Note that the discriminant is a homogeneous function of weight $n(n - 1)$ in the roots of the degree n polynomial f.

The powerful theorem of Chebotarev states that the proportion of partitions of n induced by the degrees of irreducible factors of f modulo successive primes tends to the proportion of elements occurring in $\text{Gal}_Q f$ with that partition as their shape.

Unfortunately, this result cannot be used even though effective bounds have been calculated (in terms of the discriminant) so as to ensure that all shapes in $\text{Gal}_Q f$ have occurred; this is because the bounds are too large to be practical. In practice, although the proportions give one some idea of the likely $\text{Gal}_Q f$, one uses the weaker result of the Frobenius element, merely noting occurrences and not frequencies. This part of the computation is driven by tables of groups and the shapes of their elements (see [BM]). The computation of shapes gives most information when $\text{Gal}_Q f$ contains elements of many shapes, as occurs in the symmetric and alternating groups. The minimal yield occurs with cyclic groups of prime degree. A current stopping criterion for C is that the set of possible groups for $\text{Gal}_Q f$ has not changed while factoring modulo the last n primes. More work may yield a better criterion. Further ideas are needed to extract information from the bad primes.

If $\text{Gal}_Q f$ has not been determined, we move to O and compute the orbit lengths of $\text{Gal}_Q f$ acting as described above on various r-sets. If $\text{Gal}_Q f$ remains undetermined, we resort to ad hoc methods to complete the job.

It is our intention to test these techniques using several algebraic systems and to compare them with the Galois group program as a benchmark. The current implementation is in the multilength integer language ALGEB, which was developed by David Ford [F] for use on the VAX (under VMS) and the IBM-PC.

Determining $\text{Gal}_Q f$ from f might be called the (easy) main problem in Galois theory. The (hard) inverse problem is to determine which groups can occur as $\text{Gal}_Q f$. Until recently the central result here was that of Shafarevic, who showed that all solvable groups occur. His proof is complicated and it would seem that a simpler proof should be sought. Thompson [T] has produced a sufficient criterion for realizability which has been used to show that the monster simple sporadic group of order approximately 10^{54} is realizable as a Galois group over the rationals. It can also be used to prove that the sporadic group M_{12} is so realizable. Unfortunately, Thompson's criterion in its present form is not necessary, as exemplified by M_{11}, which has been shown by Matzat [M1, M2] to be realizable over the rationals. To show all groups are realizable as $\text{Gal}_Q f$ for some f it is necessary to show that all simple groups are realizable and that the extension problem can be solved.

Computation of Galois Groups 149

3. AN EXAMPLE

An explicit one-parameter family has been constructed by Matzat [M2] which contains polynomials over the rationals with Galois group M_{12}. One such is the monic polynomial f with coefficients 1, 10, 1050, 83 700, 888 975, 3 645 000, -10 570 500, -107 163 000, 100 875 375, 1 131 772 500, -329 614 375, 1 328 602 300, and 332.150 625. As in [EFM], the Galois group preserves a Steiner system; here it is $S(5,6,12)$, namely, a set of 6-ads chosen from a set of size 12 (which we take to be the finite field $F_{11} \cup \{\infty\}$ such that each 5-ad lies in a unique 6-ad.

Since the discriminant and the degree shapes derived from factoring modulo p for (say) p = 7, 13, 17, 19, 31, 37 tell us that $M_{12} \subseteq \text{Gal}_Q f \subseteq A_{12}$, what is needed is to prove that $\text{Gal}_Q f \neq A_{12}$ since M_{12} is maximal in A_{12}. One way to do this is to prove that $\text{Gal}_Q f$ is not transitive on sets of the roots taken six at a time. An alternative is to prove that $\text{Gal}_Q f$ does indeed satisfy the definitive property for M_{12}, namely, that it is the maximal transitive group of degree 12 preserving $S(5,6,12)$.

The following gives us $S(5,6,12)$:

The elements of $F_{11} \cup \{\infty\}$ are identified with the corresponding subspaces of F_{11}^2 and we apply the linear transformation T_b, defined by

$$T_b: \begin{matrix} 1 & 0 \\ b^{-2} & 1 \end{matrix}$$

to obtain the 6-ads as images of the six squares:

T_∞:	0	1	4	9	5	3
T_1:	0	6	3	2	10	9
T_2:	0	3	2	7	1	8
T_3:	0	2	7	10	4	5
T_4:	0	10	1	4	8	6
T_5:	0	9	8	5	6	7

These and their complements, together with their images under the 11-cycle $i \to i + 1$ (fixing ∞), are the 132 6-ads of $S(5,6,12)$. When we identify the roots of f, given here to the nearest Gaussian integer, as

$$x_0 = -21 - 3i \quad x_4 = \bar{x}_1 \quad x_8 = 0$$
$$x_1 = 3 - i \quad x_5 = -i \quad x_9 = -5$$
$$x_2 = -10 - 4i \quad x_6 = \bar{x}_0 \quad x_{10} = -16$$
$$x_3 = \bar{x}_2 \quad x_7 = \bar{x}_5 \quad x_\infty = -23$$

we find that the value of the sum of the products of roots with subscripts in a 6-ad has an (approximate) numerical value of -7 508 700. A proof that this value is exact establishes that $\text{Gal}_Q f = M_{12}$.

Note added in proof: A program using the techniques described above to compute $\text{Gal}_Q f$ from f for deg f $\leq$ 7 has been incorporated into the MAPLE library.

REFERENCES

[BM] Butler, G., and McKay, J, The transitive groups of degree up to eleven, Comm. Algebra *11* (1983), 863-911.

[C] Cameron, P., Transitivity of permutation groups on unordered sets, Math. Z. *148* (1976), 127-139.

[EFM] Erbach, D. W., Fischer, J., and McKay, J., Polynomials with PSL(2,7) as Galois groups, J. Number Theory *11* (1979), 69-75.

[F] Ford, D., Computation of the maximal order in a Dedekind domain, Ph.D. thesis, Ohio State University, 1970.

[KM] Kolesova, G., and McKay, J., Practical strategies for computing Galois groups, in *Computational Group Theory*, ed. M. D. Atkinson, Academic Press, New York, 1984, pp. 297-299.

[M1] Matzat, B. H., Konstruktion von Zahl - und Funktionenkörpern mit vorgegebener Galoisgruppe, Crelle *349* (1983), 179-220.

[M2] Matzat, B. H., Realising M_{11} and M_{12} as Galois groups over Q, preprint.

[Mc] McKay, J., Some remarks on computing Galois groups, SIAM J. Comput. *8* (1979), 344-347.

[McR] McKay, J., and Regener, E., Actions of permutation groups on r-sets, Comm. Algebra *13* (1985), 619-630.

[S1] Soicher, L., The computation of Galois groups, Master's thesis, Concordia University, Montreal, Quebec, Canada, 1981.

[S2] Soicher, L., An algorithm for computing Galois groups, in *Computational Group Theory*, ed. M. D. Atkinson, Academic Press, New York, 1984, pp. 291-296.

[SMc] Soicher, L., and McKay, J., Computing Galois groups over the rationals, J. Number Theory *20* (1985), 273-281.

[T] Thompson, J. G., Some finite groups which appear as Gal L/K, where K is contained in $Q(\mu_n)$, J. Algebra *89* (1984), 437-449.

[vdW] van der Waerden, B. L., *Algebra*, Vol. 1, Frederick Ungar, New York, 1970.

On the Use of SCRATCHPAD in the Construction of Convolution Algorithms

LOUIS AUSLANDER   Graduate Center of the City University of New York, New York, New York

ALLAN J. SILBERGER   Cleveland State University, Cleveland, Ohio, and IBM Thomas J. Watson Research Center, Yorktown Heights, New York

1. INTRODUCTION

The purpose of this chapter is to describe an interesting application of SCRATCHPAD and SCRATCHPAD-like languages to the construction of algorithms used in digital signal processing. We discuss only the construction of algorithms for cyclic convolution. The best published results appear in the well-known paper of Agarwal and Cooley [1]. Our methods involve using SCRATCHPAD to compute noncyclic convolutions in certain algebraic extension fields of the rational number field.

The authors, in their study of digital signal processing, have been attempting to construct algorithms for computing cyclic convolutions which satisfy a certain stability condition and, among algorithms that satisfy this stability condition, require a minimum number of multiplications. These algorithms, except for the very simplest cases, are certainly not minimal algorithms. Thus the theory of minimal algorithms, presented in the work of Winograd [2] does not provide a theory for their construction. We still use the Chinese remainder theorem to construct these algorithms. What appears surprising, however, is that, in contrast to the case of minimal algorithms, we seem to need the Chinese remainder theorem over algebraic extension fields of the rational number field.

We present a detailed discussion of the construction of a stable algorithm for cyclic convolution on nine points. We begin by discussing the simpler problem of constructing an algorithm for cyclic convolution on three points. For this case minimal algorithms exist which are also stable. Although the result we discuss for three-point convolution is part of the theory of minimal algorithms, our discussion offers the reader some insight into the approach we take toward the more complicated nine-point case. Moreover, we use the three-point algorithm in constructing an algorithm for cyclic convolution on nine points.

Although S. Winograd is not a co-author of this paper, our understanding of the algorithms discussed here developed during discussions with him. Winograd was the first to study stable algorithms. He was aware before we were of the existence of a stable algorithm on nine points like the one we construct here.

2. COMPUTING CYCLIC CONVOLUTION ON THREE POINTS

We wish to compute the product

$$X(u)Y(u) = (x_0 + x_1 u + x_2 u^2)(y_0 + y_1 u + y_2 u^2)$$

in the algebra $Q[u,x,y]/(u^3 - 1)$, where u and the coefficients $x = (x_0, \ldots, x_2)$ and $y = (y_0, \ldots, y_2)$ are regarded as indeterminates. In counting multiplications we shall follow the conventions in Ref. 2. This amounts to counting as a multiplication any product in which each factor is an expression involving indeterminates. In the case of three-point cyclic convolution we know from Ref. 2 that every algorithm that effects this computation must include at least four multiplications.

A multiplicatively minimal algorithm satisfying our stability condition may be constructed as follows. We first observe that in $Q[u]/(u^3 - 1)$,

$$e_1 = \frac{1}{3}(1 + u + u^2)$$

and

$$e_2 = \frac{1}{3}(2 - u - u^2)$$

are orthogonal idempotents such that

$$1 = e_1 + e_2$$

Thus
$$X(u)Y(u) = e_1 X(u)Y(u) + e_2 X(u)Y(u)$$
$$= e_1(e_1 X(u))(e_1 Y(u)) + e_2(e_2 X(u))(e_2 Y(u))$$

Letting
$$X_j(u) = e_j X(u)$$

and
$$Y_j(u) = e_j Y(u) \quad j = 1, 2$$

we have
$$X(u)Y(u) = e_1 X_1(u) Y_1(u) + e_2 X_2(u) Y_2(u)$$

Since
$$ue_1 = e_1$$

and
$$u^2 e_2 = -(u + 1)e_2$$

it follows that
$$e_1 X_1(u) Y_1(u) = e_1(x_0 + x_1 + x_2)(y_0 + y_1 + y_2)$$

and that
$$e_2 X_2(u) Y_2(u) = e_2[(x_0 - x_2) + (x_1 - x_2)u][(y_0 - y_2) + (y_1 - y_2)u]$$
$$= e_2\{[(x_0 - x_2)(y_0 - y_2) - (x_1 - x_2)(y_1 - y_2)]$$
$$+ [(x_0 - x_2)(y_0 - y_2) - (x_0 - x_1)(y_0 - y_1)]u\}$$

From this we see that the product $X(u)Y(u)$ may be computed by first computing the multiplications

$$m_0 = (x_0 + x_1 + x_2)(y_0 + y_1 + y_2)$$
$$m_1 = (x_0 - x_2)(y_0 - y_2)$$
$$m_2 = (x_0 - x_1)(y_0 - y_1)$$

and
$$m_3 = (x_1 - x_2)(y_1 - y_2)$$

and then noting that

$$X(u)Y(u) = \frac{1}{3}(1 + u + u^2)m_0 + \frac{1}{3}(2 - u - u^2)[(m_1 - m_3) + (m_1 - m_2)u]$$

Expanding and collecting terms, we may write $X(u)Y(u)$ as a polynomial in u whose coefficients are linear forms in the variables $m_0, \ldots, m_3$. Using matrix notation, we write

$$X(u)Y(u) = \frac{1}{3}[1 \ u \ u^2] \begin{bmatrix} 1 & 1 & 1 & -2 \\ 1 & 1 & -2 & 1 \\ 1 & -2 & 1 & 1 \end{bmatrix} \begin{bmatrix} m_0 \\ m_1 \\ m_2 \\ m_3 \end{bmatrix}$$

$$= \frac{1}{3}[1 \ u \ u^2] \begin{bmatrix} 1 & 1 & 1 & -2 \\ 1 & 1 & -2 & 1 \\ 1 & -2 & 1 & 1 \end{bmatrix} \left(\begin{bmatrix} 1 & 1 & 1 \\ 1 & 0 & -1 \\ 1 & -1 & 0 \\ 0 & 1 & -1 \end{bmatrix} \begin{bmatrix} x_0 \\ x_1 \\ x_2 \end{bmatrix} \right.$$

$$\left. \times \begin{bmatrix} 1 & 1 & 1 \\ 1 & 0 & -1 \\ 1 & -1 & 0 \\ 0 & 1 & -1 \end{bmatrix} \begin{bmatrix} y_0 \\ y_1 \\ y_3 \end{bmatrix} \right)$$

(The "×" denotes componentwise multiplication.)

For our purposes here an algorithm will be a matrix product of the form

$A(Bx \times By)$

where A and B are matrices of constants and x and y are vectors of variables. The part within the parentheses represents the multiplication steps. We shall call an algorithm stable when the matrix B has no entries different from 1, 0, or -1.

Besides being stable, the algorithm above for computing three-point cyclic convolution is also minimal in the sense that no algorithm can exist such that the matrix B has fewer than four rows. The algorithm that we shall soon describe for computing cyclic convolution on nine points will be stable, but it will not be multiplicatively minimal. However, we conjecture that it is multiplicatively minimal in the class of stable algorithms.

Let us remark that even though it is important to do so, we do not attempt to represent the algorithms we construct as sequences of field operations nor even to estimate the number of additions needed in computing with these algorithms. Agarwal and Cooley give this information in Ref. 1 for the algorithms they construct.

Before turning to the nine-point case, we rederive the three-point algorithm above via an approach which is closer to the one we shall take in the nine-point case. Let ω be a primitive cube root of unity and consider the algebra $Q[\omega,u,x,y]/(u^3 - 1)$. In this algebra we may write 1 as the sum of the three orthogonal idempotents

$$f_1 = e_1 = \frac{1}{3}(1 + u + u^2)$$

$$f_2 = \frac{1}{3}(1 + \omega^2 u + \omega u^2)$$

and

$$f_3 = \frac{1}{3}(1 + \omega u + \omega^2 u^2)$$

We obtain

$$X(u)Y(u) = f_1 \mu_1 + f_2 \mu_\omega + f_3 \mu_{\bar\omega}$$

where

$$\mu_1 = (x_0 + x_1 + x_2)(y_0 + y_1 + y_2)$$
$$\mu_\omega = (x_0 + \omega x_1 + \omega^2 x_2)(y_0 + \omega y_1 + \omega^2 y_2)$$

and

$$\mu_{\bar\omega} = (x_0 + \omega^2 x_1 + \omega x_2)y_0 + \omega^2 y_1 + \omega y_2)$$

From this formula follows by inspection the formula

$$X(u)Y(u) = \frac{1}{3}(\mu_1 + \mu_\omega + \mu_{\bar\omega}) + \frac{1}{3}(\mu_1 + \omega^2 \mu_\omega + \omega \mu_{\bar\omega})u$$
$$+ \frac{1}{3}(\mu_1 + \omega \mu_\omega + \omega^2 \mu_{\bar\omega})u^2$$

To recover our rational algorithm we replace ω by the companion matrix

$$\begin{bmatrix} 0 & -1 \\ 1 & -1 \end{bmatrix}$$

and check that

$$\mu_1 = \begin{bmatrix} m_0 & 0 \\ 0 & m_0 \end{bmatrix}$$

$$\mu_\omega = \begin{bmatrix} m_1 - m_3 & m_2 - m_1 \\ m_1 - m_2 & m_2 - m_3 \end{bmatrix}$$

and

$$\mu_{\bar{\omega}} = \begin{bmatrix} m_2 - m_3 & m_1 - m_2 \\ m_2 - m_1 & m_1 - m_3 \end{bmatrix}$$

We obtain our first algorithm by substituting these expressions into the formula expressing $X(u)Y(u)$ in terms of μ_1, μ_ω, and $\mu_{\bar{\omega}}$.

3. CYCLIC CONVOLUTION ON NINE POINTS

To construct an algorithm to compute cyclic convolution on nine points we apply the second of the approaches above to three-point cyclic convolution to computing multiplication in the algebra $Q[u,x,y]/(u^9 - 1)$, where, from now on, $x = (x_0,\ldots,x_8)$ and $y = (y_0,\ldots,y_8)$. Since

$$e_1 = \frac{1}{3}(1 + u^3 + u^6)$$

and

$$e_2 = \frac{1}{3}(2 - u^3 - u^6)$$

are orthogonal idempotents whose sum is one in the algebra $Q[u]/(u^9 - 1)$, it follows that if $X(u) = \sum_{i=0}^{8} x_i u^i$ and $Y(u) = \sum_{i=0}^{8} y_i u^i$, then

$$X(u)Y(u) = e_1 X_1(u)Y_1(u) + e_2 X_2(u)Y_2(u)$$

where

$$X_1(u) = \sum_{i=0}^{2} (x_i + x_{i+3} + x_{i+6})u^i$$

$$Y_1(u) = \sum_{i=0}^{2} (y_i + y_{i+3} + y_{i+6})u^i$$

$$X_2(u) = \sum_{i=0}^{2} [(x_i - x_{i+6})u^i + (x_{i+3} - x_{i+6})u^{i+3}]$$

and

$$Y_2(u) = \sum_{i=0}^{2} [(y_i - y_{i+6})u^i + (y_{i+3} - y_{i+6})u^{i+3}]$$

The product $X_1(u)Y_1(u)$ may be computed by using the algorithm we constructed for three-point convolution. We define

$$m_0 = (x_0 + \cdots + x_8)(y_0 + \cdots + y_8)$$
$$m_1 = (x_0 - x_2 + x_3 - x_5 + x_6 - x_8)(y_0 - y_2 + y_3 - y_5 + y_6 - y_8)$$
$$m_2 = (x_0 - x_1 + x_3 - x_4 + x_6 - x_7)(y_0 - y_1 + y_3 - y_5 + y_6 - y_7)$$

and

$$m_3 = (x_1 - x_2 + x_4 - x_5 + x_7 - x_8)(y_1 - y_2 + y_4 - y_5 + y_7 - y_8)$$

The algorithm for three-point convolution given previously implies that

$$e_1 X_1(u) Y_1(u) = \frac{e_1}{3} \begin{bmatrix} 1 & u & u^2 \end{bmatrix} \begin{bmatrix} 1 & 1 & 1 & -2 \\ 1 & 1 & -2 & 1 \\ 1 & -2 & 1 & 1 \end{bmatrix} \begin{bmatrix} m_0 \\ m_1 \\ m_2 \\ m_3 \end{bmatrix}$$

We turn to the computation of $e_2 X_2(u) Y_2(u)$. Over $Q[\omega]$ we may write

$$e_1 = f_1$$

and

$$e_2 = f_2 + f_3$$

where

$$f_2 = \frac{1}{3}(1 + \omega^2 u^3 + \omega u^6)$$

and

$$f_3 = \frac{1}{3}(1 + \omega u^3 + \omega^2 u^6)$$

Therefore,

$$e_2 X_2(u) Y_2(u) = f_2(\xi_0 + \xi_1 u + \xi_2 u^2)(\eta_0 + \eta_1 u + \eta_2 u^2)$$
$$+ f_3(\bar{\xi}_0 + \bar{\xi}_1 u + \bar{\xi}_2 u^2)(\bar{\eta}_0 + \bar{\eta}_1 u + \bar{\eta}_2 u^2)$$

Here the bar denotes conjugation in the field $Q[\omega]$,

$$\xi_i = (x_i - x_{i+6}) + (x_{i+3} - x_{i+6})\omega$$

and

$$\eta_i = (y_i - y_{i+6}) + (y_{i+3} - y_{i+6})\omega$$

$i = 0, 1, 2$. The product

$$(\xi_0 + \xi_1 u + \xi_2 u^2)(\eta_0 + \eta_1 u + \eta_2 u^2)$$

must be computed in $Q[\omega,u,x,y]/(u^3 - \omega)$. The second product in $Q[\omega,u,x,y]/(u^3 - \bar{\omega})$, is just the conjugate of the first, so we have to compute only the first.

We use SCRATCHPAD to compute this product as follows. First we define the "reducing polynomial"

$$F(u) = u(u - 1)(u - \omega)(u - \bar{\omega})$$

Then we set

$$\mu_0 = \xi_0 \eta_0$$
$$\mu_1 = (\xi_0 + \xi_1 + \xi_2)(\eta_0 + \eta_1 + \eta_2)$$
$$\mu_\omega = (\xi_0 + \xi_1 \omega + \xi_2 \omega^2)(\eta_0 + \eta_1 \omega + \eta_2 \omega^2)$$
$$\mu_{\bar{\omega}} = (\xi_0 + \xi_1 \omega^2 + \xi_2 \omega)(\eta_0 + \eta_1 \omega^2 + \eta_2 \omega)$$

and

$$\mu_\infty = \xi_2 \eta_2$$

By the Chinese remainder theorem, the product in $Q[\omega,u,x,y]$ may be computed as

$$(\xi_0 + \xi_1 u + \xi_2 u^2)(\eta_0 + \eta_1 u + \eta_2 u^2) = -\frac{F(u)}{u}\mu_0 + \frac{1}{3}\frac{F(u)}{u - 1}\mu_1 + \frac{1}{3}\frac{F(u)}{u - \omega}\mu_\omega$$
$$+ \frac{1}{3}\frac{F(u)}{u - \bar{\omega}}\mu_{\bar{\omega}} + \mu_\infty F(u)$$

Reducing this product mod $u^3 - \omega$ and defining, for the reduced product,

$$\zeta_0 + \zeta_1 u + \zeta_2 u^2 = f_2(\xi_0 + \xi_1 u + \xi_2 u^2)(\eta_0 + \eta_1 u + \eta_2 u^2)$$

we find that

$$3\zeta_0 = 3\mu_0 + \omega(-3\mu_0 + \mu_1 + \mu_\omega + \mu_{\bar{\omega}})$$
$$3\zeta_1 = (\mu_1 - \mu_\omega - 3\mu_\infty) + \omega(-\mu_\omega + \mu_{\bar{\omega}} + 3\mu_\infty)$$

and

$$3\zeta_2 = (\mu_1 - \mu_{\bar{\omega}}) + \omega(\mu_\omega - \mu_{\bar{\omega}})$$

With SCRATCHPAD this computation with the Chinese remainder theorem is formal; in the present case one can easily compute $\zeta_0, \ldots, \zeta_2$ by hand. The conjugate product $\bar{\zeta}_0 + \bar{\zeta}_1 u + \bar{\zeta}_2 u^2$ comes for free. With $Z_1 = f_1 X_1(u) Y_1(u)$ and $Z_2 = \zeta_0 + \zeta_1 u + \zeta_2 u^2$, we obtain

$$X(u)Y(u) = \frac{1}{3}(Z_1 + Z_2 + \bar{Z}_2) + \frac{1}{3}(Z_1 + \omega^2 Z_2 + \omega\bar{Z}_2)u^3$$
$$+ \frac{1}{3}(Z_1 + \omega Z_2 + \omega^2 \bar{Z}_2)u^6$$

We use SCRATCHPAD to combine the terms above and replace each multiplication in $Q[\omega]$ by a set of three rational multiplications derived from the three-point algorithm. In the Appendix we give the record of our SCRATCHPAD "conversation," in which the algorithm matrices are computed.

We remark that our algorithm will compute cyclic convolution on nine points in 19 multiplications. Four are needed to perform the three-point convolution part of the computation. Fifteen rational multiplications compute the product in $Q[u,x,y]/(u^6 + u^3 + 1)$.

The multiplication in $Q[u,x,y]/(u^6 + u^3 + 1)$ may be computed using the Chinese remainder theorem over the rationals. In this case, a minimal algorithm for computing the product of two quintics requires only 11 multiplications. However, a minimal algorithm which will satisfy our stability condition requires 16 multiplications. This follows from the fact that projecting the octic $X(u)$ on $Q[u,x,y]/(u^6 + u^3 + 1)$ produces a quintic of the form

$$\sum_{i=0}^{2} [(x_i - x_{i+6})u^i + (x_{i+3} - x_{i+6})u^{k+3}]$$

To multiply two quintics of this form via the Chinese remainder theorem, one cannot use the polynomial $u - 1$ as a factor of the reducing polynomial over the rationals as this results in a matrix with entries other than 1, 0, or -1. The reducing polynomial over the rationals which results in a multiplicatively minimal stable algorithm for multiplying our quintics is

$$G(u) = u^2(u + 1)(u^2 + 1)(u^2 + u + 1)(u^2 - u + 1)$$

The choice leads to 16 multiplications.

The algorithm in Ref. 1 computes the same product of quintics in 18 multiplications by an iterative method. This algorithm is also stable.

APPENDIX

The following SCRATCHPAD conversation was printed from the spool file of our terminal display. Comments have been included to assist the reader in following the calculation of the algorithm matrices. These programs have been adapted from SCRATCHPAD programs of J. W. Cooley, whom the second author would like to thank for his patience in leading that author through his first programming experience.

```
ASILBGR CONSOLE A0 dated   84/04/14 16:12:46 ..... Page 1


SCRATCHPAD
)read cyc9c
"compute nine point cyclic convolution by making use of
the field of cube roots of one"
algnum=w**2+w+1
```

$$\text{ALGNUM} = W^2 + W + 1$$
ALGEBRAIC NUMBER EXTENSION

```
p=u**9-1
```

$$P = U^9 - 1$$

```
factor(p)
```

FACTORIZATION OVER $W^2 + W + 1 = 0$

$$(1) \quad (U^3 + W + 1)(U^3 - W)(U - 1)(U + W + 1)(U - W)$$

```
pv='parts(ws)
```

$$PV = (U^3 + W + 1, U^3 - W, U - 1, U + W + 1, U - W)$$

```
format=(a,p(u))
"factors for computing with the crt"
fv='(u,u-1,u-w,u-w**2)
```

$$FV = (U, U - 1, U - W, U + (W + 1))$$

```
"the reducing polynomial"
f='prod<j=1;4>(fv.j)
```

$$F = U^4 - U$$

```
"compute the idempotents for the crt decomposition"
t<j>='f/fv.j for j in (1,...,4)
```

$$T_1 = U^3 - 1$$

$$T_2 = U^3 + U^2 + U$$

$$T_3 = U^3 + U^2 W + U(-W - 1)$$

ASILBGR CONSOLE A0 dated 84/04/14 16:12:46 ..... Page 2

$$T_4 = U + U(-W-1) + U*W$$

s<j>='pinvmod(t<j>,fv.j,u) for j in (1,...,4)

$$S_1 = -1$$

$$S_2 = 1/3$$

$$S_3 = 1/3$$

$$S_4 = 1/3$$

e<j>='t<j>*s<j> for j in (1,...,4)

$$E_1 = -U^3 + 1$$

$$E_2 = (1/3)*U^3 + (1/3)*U^2 + (1/3)*U$$

$$E_3 = (1/3)*U^3 + U^2\left(\frac{W}{3}\right) + U\left(\frac{-W-1}{3}\right)$$

$$E_4 = (1/3)*U^3 + U^2\left(\frac{-W-1}{3}\right) + U\left(\frac{W}{3}\right)$$

"express the projections of the product polynomial in terms of the multiplications over the field of cube roots of one"

z<;5>=m<2>

$$Z^5 = M_2$$

z<;4>=m.<1>

$$Z^4 = M_1$$

z<;3>=m<0>

$$Z^3 = M$$

ASILBGR CONSOLE A0 dated   84/04/14 16:12:46 ..... Page 3

$$0$$

z<;2,1>=m<3>

$$Z^{2,1} = M_3$$

z<;2,2>=m<4>

$$Z^{2,2} = M_4$$

z<;2,3>=m<5>

$$Z^{2,3} = M_5$$

z<;2,4>=m<6>

$$Z^{2,4} = M_6$$

z<;2,5>=m<7>

$$Z^{2,5} = M_7$$

z<;1,1>=m<8>

$$Z^{1,1} = M_8$$

z<;1,2>=m<9>

$$Z^{1,2} = M_9$$

z<;1,3>=m<10>

$$Z^{1,3} = M_{10}$$

z<;1,4>=m<11>

$$Z^{1,4} = M_{11}$$

z<;1,5>=m<12>

$$1,5$$

Use of SCRATCHPAD in Algorithms   163

ASILBGR CONSOLE A0 dated 84/04/14 16:12:46 ..... Page 4

$$Z = M_{12}$$

"compute with the crt in the field of cube roots of one and the conjugate calculation"
sum<j=1;4>(z<;1,j>*e<j>)+z<;1,5>*f

(2)

$$U^4 M_{12} + U^3 \left(\frac{M_{11} + M_{10} + M_9 - 3M_8}{3}\right)$$
$$+ U^2 \left(\frac{-W*M_{11} + W*M_{10} - M_{11} + M_9}{3}\right)$$
$$+ U\left(\frac{W*M_{11} - W*M_{10} - 3M_{12} - M_{10} + M_9}{3}\right) + M_8$$

z<;1>=`remainder(numer(ws),u**3+w+1,u)/denom(ws)

$$Z^1$$

$$= U^2 \left(\frac{-W*M_{11} + W*M_{10} - M_{11} + M_9}{3}\right)$$
$$+ U\left(\frac{-3W*M_{12} + W*M_{11} - W*M_{10} - 6M_{12} - M_{10} + M_9}{3}\right)$$
$$+ \frac{-W*M_{11} - W*M_{10} - W*M_9 + 3W*M_8 - M_{11} - M_{10} - M_9 + 6M_8}{3}$$

sum<j=1;4>(z<;2,j>*e<j>)+z<;2,5>*f

(3)

$$U^4 M_7 + U^3 \left(\frac{M_6 + M_5 + M_4 - 3M_3}{3}\right)$$
$$+ U^2 \left(\frac{-W*M_6 + W*M_5 - M_6 + M_4}{3}\right)$$
$$+$$

ASILBGR CONSOLE A0 dated 84/04/14 16:12:46 ..... Page 5

$$U\left(\frac{W*M_6 - W*M_5 - 3M_7 - M_5 + M_4}{3}\right) + \frac{M}{3}$$

z<;2>='remainder(numer(ws),u**3-w,u)/denom(ws)

$$Z^2 =$$

$$U^2\left(\frac{-W*M_6 + W*M_5 - M_6 + M_4}{3}\right)$$

$$+ U\left(\frac{3W*M_7 + W*M_6 - W*M_5 - 3M_7 - M_5 + M_4}{3}\right)$$

$$+ \frac{W*M_6 + W*M_5 + W*M_4 - 3W*M_3 + 3M_3}{3}$$

"use the crt to compute the linear forms in the multiplication
steps over the cube roots of one which are coefficients of
the product polynomial"
v<j>='p/pv.j for j in (1,...,5)

$$V_1 = U^6 + U^3(-W - 1) + W$$

$$V_2 = U^6 + U^3 W + (-W - 1)$$

$$V_3 = U^8 + U^7 + U^6 + U^5 + U^4 + U^3 + U^2 + U + 1$$

$$V_4 =$$
$$U^8 + U^7(-W - 1) + U^6 W + U^5 + U^4(-W - 1) + U^3 W + U^2$$
$$+ U(-W - 1) + W$$

$$V_5 =$$
$$\quad 8 \quad 7 \quad 6 \qquad 5 \quad 4 \quad 3 \qquad 2$$

Use of SCRATCHPAD in Algorithms

ASILBGR CONSOLE A0 dated 84/04/14 16:12:46 ..... Page 6

$$U + UW + U(-W-1) + U + UW + U(-W-1) + U + U*W$$
$$+ -W - 1$$

w<j>='pinvmod(v<j>,pv.j,u) for j in (1,...,5)

$$W_1 = \frac{1}{3W}$$

$$W_2 = \frac{-1}{3W + 3}$$

$$W_3 = 1/9$$

$$W_4 = \frac{1}{9W}$$

$$W_5 = \frac{-1}{9W + 9}$$

x<j>='v<j>*w<j> for j in (1,...,5)

$$X_1 = U^6 \left(\frac{1}{3W}\right) + U^3 \left(\frac{-W-1}{3W}\right) + 1/3$$

$$X_2 = U^6 \left(\frac{-1}{3W+3}\right) + U^3 \left(\frac{-W}{3W+3}\right) + \frac{W+1}{3W+3}$$

$$X_3 = (1/9)*U^8 + (1/9)*U^7 + (1/9)*U^6 + (1/9)*U^5 + (1/9)*U^4$$
$$+ (1/9)*U^3 + (1/9)*U^2 + (1/9)*U + 1/9$$

$$X_4 =$$
$$U^8 \left(\frac{1}{9W}\right) + U^7 \left(\frac{-W-1}{9W}\right) + (1/9)*U^6 + U^5 \left(\frac{1}{9W}\right)$$
$$+ U^4 \left(\frac{-W-1}{9W}\right) \quad U^3 \quad U^2 \quad U^1 \quad \left(\frac{-W-1}{}\right)$$

```
ASILBGR CONSOLE A0 dated  84/04/14 16:12:46 ..... Page 7


                   U (---------) + (1/9)*U  + U (----) + U(---------) + 1/9
                        9W                       9W         9W


         X
          5
           =
                 8    -1            7   - W           6  W + 1       5    -1
                U (--------) + U (--------) + U (--------) + U (--------)
                     9W + 9            9W + 9            9W + 9           9W + 9
             +
                 4    - W           3  W + 1         2   -1             - W
                U (--------) + U (--------) + U (--------) + U(--------)
                     9W + 9            9W + 9            9W + 9           9W + 9
             +
                  W + 1
                 --------
                  9W + 9

sum<j=1;5>(z<;j>*x<j>)*9

         (4)
                     8
                    U
                    *
                         W*M   - W*M  - W*M  + W*M  + W*M  - W*M  + M   - M  + M
                            11      9      5      4      2      1     10     9    6
                      +
                         - M  - M  + M
                            5    1    0
                 +
                     7
                    U
                    *
                         6W*M  + W*M   - W*M  - 6W*M  - W*M  + W*M  - W*M
                             12     10      9       7      6      4      2
                      +
                         W*M  + 3M   + M   - M  - 3M  - M  + M  - M  + M
                            1    12     11      9     7     6     5     2    0
                 +
                     6
                    U
                    *
                         W*M   + W*M   + W*M  - 6W*M  - W*M  - W*M  - W*M
                            11      10      9       8      6      5      4
                      +
                         6W*M  - 3M  - M  - M  - M  + 3M  + M  + M  + M
                             3     8    6    5    4     3    2    1    0
                 +
                     5
                    U
                    *
                         - W*M  + W*M  + W*M  - W*M  + W*M  - W*M  + M   - M
                              10      9      6      4      2      1     11     10
                      +
                         M  - M  - M  + M
                          5    4    1    0
                 +
```

Use of SCRATCHPAD in Algorithms

$$
\begin{aligned}
&U^4 \\
&\quad * \quad -3W*M_{12} - W*M_{11} + W*M_9 + 3W*M_7 + W*M_5 - W*M_4 - W*M_2 \\
&\quad + \quad W*M_1 + 3M_{12} - M_{11} + M_{10} + 6M_7 + M_6 - M_4 - M_2 + M_0 \\
&+ \\
&U^3 \\
&\quad * \quad 3W*M_8 - 3W*M_3 + M_{11} + M_{10} + M_9 - 3M_8 + M_6 + M_5 + M_4 \\
&\quad + \quad -6M_3 + M_2 + M_1 + M_0 \\
&+ \\
&U^2 \\
&\quad * \quad -W*M_{11} + W*M_{10} - W*M_6 + W*M_5 + W*M_2 - W*M_1 - M_{11} + M_9 \\
&\quad + \quad -M_6 + M_4 - M_1 + M_0 \\
&+ \\
&U \\
&\quad * \quad -3W*M_{12} + W*M_{11} - W*M_{10} + 3W*M_7 + W*M_6 - W*M_5 - W*M_2 \\
&\quad + \quad W*M_1 - 6M_{12} - M_{10} + M_9 - 3M_7 - M_5 + M_4 - M_2 + M_0 \\
&\quad + \quad -W*M_{11} - W*M_{10} - W*M_9 + 3W*M_8 + W*M_6 + W*M_5 + W*M_4 - 3W*M_3 \\
&\quad + \quad -M_{11} - M_{10} - M_9 + 6M_8 + 3M_3 + M_2 + M_1 + M_0
\end{aligned}
$$

"the vector of linear forms"
h=`covec(u,ws)`

$$
H = \Big(\; 0 \; : \;
\begin{aligned}
&-W*M_{11} - W*M_{10} - W*M_9 + 3W*M_8 + W*M_6 + W*M_5 \\
&+ \\
&W*M_4 - 3W*M_3 - M_{11} - M_{10} - M_9 + 6M_8 + 3M_3 + M_2 \\
&+ \\
&M_1 + M_0
\end{aligned}
$$

ASILBGR CONSOLE A0 dated 84/04/14 16:12:46 ..... Page 9

$$\begin{aligned}
&- 3W^*M_{12} + W^*M_{11} - W^*M_{10} + 3W^*M_{7} + W^*M_{6} - W^*M_{5} - W^*M_{2} \\
&+ W^*M_{1} - 6M_{12} - M_{10} + M_{9} - 3M_{7} - M_{5} + M_{4} - M_{2} + M_{0} \\
&,\\
&- W^*M_{11} + W^*M_{10} - W^*M_{6} + W^*M_{5} + W^*M_{2} - W^*M_{1} - M_{11} \\
&+ M_{9} - M_{6} + M_{4} - M_{1} + M_{0} \\
&,\\
&3W^*M_{8} - 3W^*M_{3} + M_{11} + M_{10} + M_{9} - 3M_{8} + M_{6} + M_{5} + M_{4} \\
&+ - 6M_{3} + M_{2} + M_{1} + M_{0} \\
&,\\
&- 3W^*M_{12} - W^*M_{11} + W^*M_{9} + 3W^*M_{7} + W^*M_{5} - W^*M_{4} - W^*M_{2} \\
&+ W^*M_{1} + 3M_{12} - M_{11} + M_{10} + 6M_{7} + M_{6} - M_{4} - M_{2} + M_{0} \\
&,\\
&- W^*M_{10} + W^*M_{9} + W^*M_{6} - W^*M_{4} + W^*M_{2} - W^*M_{1} + M_{11} \\
&+ - M_{10} + M_{5} - M_{4} - M_{1} + M_{0} \\
&,\\
&W^*M_{11} + W^*M_{10} + W^*M_{9} - 6W^*M_{8} - W^*M_{6} - W^*M_{5} - W^*M_{4} \\
&+ 6W^*M_{3} - 3M_{8} - M_{6} - M_{5} - M_{4} + 3M_{3} + M_{2} + M_{1} + M_{0} \\
&,\\
&6W^*M_{12} + W^*M_{10} - W^*M_{9} - 6W^*M_{7} - W^*M_{6} + W^*M_{4} - W^*M_{2} \\
&+ W^*M_{1} + 3M_{12} + M_{11} - M_{9} - 3M_{7} - M_{6} + M_{5} - M_{2} + M_{0} \\
&,\\
&W^*M_{11} - W^*M_{9} - W^*M_{5} + W^*M_{4} + W^*M_{2} - W^*M_{1} + M_{10} - M_{9}
\end{aligned}$$

Use of SCRATCHPAD in Algorithms 169

ASILBGR CONSOLE A0 dated 84/04/14 16:12:46 ..... Page 10

$$+ M_6 - M_5 - M_1 + M_0 ,$$

ENDOFFILE : CYC9C

SCRATCHPAD
)read newmat9
SCRATCHPAD
"use the regular representation of the field of cube roots of one to obtain a rational algorithm for nine point cyclic convolution"
"here the m's are multiplications in the cyclotomic field and the n's are rational multiplications to be described later."
m<0>=n<0>

$$M_0 = N_0$$

m<2>=(-n<3>-n<2>*w+n<1>*(w+1))

$$M_2 = -N_3 - N_2 W + N_1 (W + 1)$$

m<1>=(-n<3>+n<2>*(w+1)-n<1>*w)

$$M_1 = -N_3 + N_2 (W + 1) - N_1 W$$

m<3>=(-n<6>-n<5>*w+n<4>*(w+1))

$$M_3 = -N_6 - N_5 W + N_4 (W + 1)$$

m<8>=(-n<6>+n<5>*(w+1)-n<4>*w)

$$M_8 = -N_6 + N_5 (W + 1) - N_4 W$$

m<4>=(-n<9>-n<8>*w+n<7>*(w+1))

$$M_4 = -N_9 - N_8 W + N_7 (W + 1)$$

m<9>=(-n<9>+n<8>*(w+1)-n<7>*w)

$$M_9 = -N_9 + N_8 (W + 1) - N_7 W$$

m<5>=(-n<12>-n<11>*w+n<10>*(w+1))

$$M_5 = -N_{12} - N_{11} W + N_{10} (W + 1)$$

m<11>=(-n<12>+n<11>*(w+1)-n<10>*w)

$$M = -N + N (W + 1) - N W$$

ASILBGR CONSOLE A0 dated 84/04/14 16:12:46 ..... Page 11

```
              11      12      11              10
m<6>=(-n<15>-n<14>*w+n<13>*(w+1))

     M  =    N    -  N  W  +  N  (W + 1)
      6      15      14         13

m<10>=(-n<15>+n<14>*(w+1)-n<13>*w)

     M  = -  N    +  N  (W + 1) -  N  W
     10      15      14            13

m<7>=(-n<18>-n<17>*w+n<16>*(w+1))

     M  = -  N    -  N  W  +  N  (W + 1)
      7      18      17         16

m<12>=(-n<18>+n<17>*(w+1)-n<16>*w)

     M  = -  N    +  N  (W + 1) -  N  W
     12      18      17            16
```

h.0

(5)

```
          N  + N  - 2N  + N  + N  - 2N  + N  + N  - 2N  - 9N
          15    14    13    12    11    10    9    8    7    6
        +
          9N  - 2N  + N  + N  + N
          4     3    2    1    0
```

h.1

(6)

```
          9N  - 9N  + N  + N  - 2N  + N  - 2N  + N  - 2N  + N
          18    16    15    14    13    12    11    10    9    8
        +
          N  + N  - 2N  + N  + N
          7    3    2    1    0
```

h.2

(7)

```
          N  - 2N  + N  + N  + N  - 2N  - 2N  + N  + N  + N  + N
          15    14    13    12    11    10    9     8    7    3    2
        +
          - 2N  + N
            1    0
```

h.3

(8)

```
          - 2N  + N  + N  - 2N  + N  + N  - 2N  + N  + N  + 9N
             15    14    13    12    11    10    9    8    7    6
        +
```

Use of SCRATCHPAD in Algorithms 171

ASILBGR CONSOLE A0 dated 84/04/14 16:12:46 ..... Page 12

$$-9N_5 - 2N_3 + N_2 + N_1 + N_0$$

h.4

(9)
$$-9N_{18} + 9N_{17} - 2N_{15} + N_{14} + N_{13} + N_{12} + N_{11} - 2N_{10} + N_9 - 2N_8$$
$$+ N_7 + N_3 - 2N_2 + N_1 + N_0$$

h.5

(10)
$$N_{15} + N_{14} - 2N_{13} - 2N_{12} + N_{11} + N_{10} + N_9 - 2N_8 + N_7 + N_3 + N_2$$
$$+ -2N_1 + N_0$$

h.6

(11)
$$N_{15} - 2N_{14} + N_{13} + N_{12} - 2N_{11} + N_{10} + N_9 - 2N_8 + N_7 + 9N_5$$
$$+ -9N_4 - 2N_3 + N_2 + N_1 + N_0$$

h.7

(12)
$$-9N_{17} + 9N_{16} + N_{15} - 2N_{14} + N_{13} - 2N_{12} + N_{11} + N_{10} + N_9 + N_8$$
$$+ 2N_7 + N_3 - 2N_2 + N_1 + N_0$$

h.8

(13)
$$-2N_{15} + N_{14} + N_{13} + N_{12} - 2N_{11} + N_{10} + N_9 + N_8 - 2N_7 + N_3$$
$$+ N_2 - 2N_1 + N_0$$

n=(n<i> for i in (0,...,18))

N= (N$_I$ FOR I IN (0,...,18))

ASILBGR CONSOLE A0 dated 84/04/14 16:12:46 ..... Page 13

"this is the matrix 9a."
coeff(n,h)

(14)
```
*1  1  1 -2  9  0 -9 -2  1  1 -2  1  1 -2  1  1  0  0  0*
*1  1 -2  1  0  0  0  1  1 -2  1 -2  1 -2  1  1 -9  0  9*
*1 -2  1  1  0  0  0  1  1 -2 -2  1  1  1 -2  1  0  0  0*
*1  1  1 -2  0 -9  9  1  1 -2  1  1 -2  1  1 -2  0  0  0*
*1  1 -2  1  0  0  0  1 -2  1 -2  1  1  1  1 -2  0  9 -9*
*1 -2  1  1  0  0  0  1 -2  1  1  1 -2 -2  1  1  0  0  0*
*1  1  1 -2 -9  9  0  1 -2  1  1 -2  1  1 -2  1  0  0  0*
*1  1 -2  1  0  0  0 -2  1  1  1  1 -2  1 -2  1  9 -9  0*
*1 -2  1  1  0  0  0 -2  1  1  1 -2  1  1  1 -2  0  0  0*
```

transpose(ws)

(15)
```
*1   1   1   1   1   1   1   1   1*
*1   1  -2   1   1  -2   1   1  -2*
*1  -2   1   1  -2   1   1  -2   1*
*-2  1   1  -2   1   1  -2   1   1*
*9   0   0   0   0   0  -9   0   0*
*0   0   0  -9   0   0   9   0   0*
*-9  0   0   9   0   0   0   0   0*
*-2  1   1   1   1   1   1  -2  -2*
*1   1   1   1  -2  -2  -2   1   1*
*1  -2  -2  -2   1   1   1   1   1*
*-2  1  -2   1  -2   1   1   1   1*
*1  -2   1   1   1   1  -2   1  -2*
*1   1   1  -2   1  -2   1  -2   1*
*-2 -2   1   1   1  -2   1   1   1*
*1   1  -2   1   1   1  -2  -2   1*
*1   1   1  -2  -2   1   1   1  -2*
*0  -9   0   0   0   0   0   9   0*
*0   0   0   0   9   0   0  -9   0*
```

ASILBGR CONSOLE A0 dated 84/04/14 16:12:46 ..... Page 14

```
        *                         *
        *0  9  0  0  -9  0  0  0  *
```

ENDOFFILE : NEWMAT9

ASILBGR CONSOLE A0 dated 84/04/14 16:35:50 ..... Page 1

"now we express the n's in terms of the coefficients of two
octic polynomials a(u) and b(u). these are the polynomials
we multiply in this program instead of x(u) and y(u)."
p<0>='sum<j=0;8>a<j>

$$P_0 = A_8 + A_7 + A_6 + A_5 + A_4 + A_3 + A_2 + A_1 + A_0$$

q<0>='sum<j=0;8>b<j>

$$Q_0 = B_8 + B_7 + B_6 + B_5 + B_4 + B_3 + B_2 + B_1 + B_0$$

n<0>=p<0>*q<0>

$$N_0 = P_0 Q_0$$

p<1>=(a<0>-a<2>+a<3>-a<5>+a<6>-a<8>)

$$P_1 = A_0 - A_2 + A_3 - A_5 + A_6 - A_8$$

q<1>=(b<0>-b<2>+b<3>-b<5>+b<6>-b<8>)

$$Q_1 = B_0 - B_2 + B_3 - B_5 + B_6 - B_8$$

n<1>=p<1>*q<1>

$$N_1 = P_1 Q_1$$

p<2>=(a<0>-a<1>+a<3>-a<4>+a<6>-a<7>)

$$P_2 = A_0 - A_1 + A_3 - A_4 + A_6 - A_7$$

q<2>=(b<0>-b<1>+b<3>-b<4>+b<6>-b<7>)

$$Q_2 = B_0 - B_1 + B_3 - B_4 + B_6 - B_7$$

n<2>=p<2>*q<2>

$$N_2 = P_2 Q_2$$

p<3>=(a<1>-a<2>+a<4>-a<5>+a<7>-a<8>)

$$P_3 = A_1 - A_2 + A_4 - A_5 + A_7 - A_8$$

q<3>=(b<1>-b<2>+b<4>-b<5>+b<7>-b<8>)

$$Q_3 = B_1 - B_2 + B_4 - B_5 + B_7 - B_8$$

n<3>=p<3>*q<3>

ASILBGR CONSOLE A0 dated 84/04/14 16:35:50 ..... Page 2

$$N_3 = P_3 Q_3$$

p<4>=(a<0>-a<6>)

$$P_4 = A_0 - A_6$$

q<4>=(b<0>-b<6>)

$$Q_4 = B_0 - B_6$$

n<4>=p<4>*q<4>

$$N_4 = P_4 Q_4$$

p<5>=(a<0>-a<3>)

$$P_5 = A_0 - A_3$$

q<5>=(b<0>-b<3>)

$$Q_5 = B_0 - B_3$$

n<5>=p<5>*q<5>

$$N_5 = P_5 Q_5$$

p<6>=(a<3>-a<6>)

$$P_6 = A_3 - A_6$$

q<6>=(b<3>-b<6>)

$$Q_6 = B_3 - B_6$$

n<6>=p<6>*q<6>

$$N_6 = P_6 Q_6$$

p<7>=(a<0>+a<1>+a<2>-a<6>-a<7>-a<8>)

$$P_7 = A_0 + A_1 + A_2 - A_6 - A_7 - A_8$$

q<7>=(b<0>+b<1>+b<2>-b<6>-b<7>-b<8>)

$$Q_7 = B_0 + B_1 + B_2 - B_6 - B_7 - B_8$$

ASILBGR CONSOLE A0 dated 84/04/14 16:35:50 ..... Page 3

n<7>=p<7>*q<7>

$$N_7 = P_7 \, Q_7$$

p<8>=(a<0>+a<1>+a<2>-a<3>-a<4>-a<5>)

$$P_8 = A_0 + A_1 + A_2 - A_3 - A_4 - A_5$$

q<8>=(b<0>+b<1>+b<2>-b<3>-b<4>-b<5>)

$$Q_8 = B_0 + B_1 + B_2 - B_3 - B_4 - B_5$$

n<8>=p<8>*q<8>

$$N_8 = P_8 \, Q_8$$

p<9>=(a<3>+a<4>+a<5>-a<6>-a<7>-a<8>)

$$P_9 = A_3 + A_4 + A_5 - A_6 - A_7 - A_8$$

q<9>=(b<3>+b<4>+b<5>-b<6>-b<7>-b<8>)

$$Q_9 = B_3 + B_4 + B_5 - B_6 - B_7 - B_8$$

n<9>=p<9>*q<9>

$$N_9 = P_9 \, Q_9$$

p<10>=(a<0>-a<2>-a<4>+a<5>-a<6>+a<7>)

$$P_{10} = A_0 - A_2 - A_4 + A_5 - A_6 + A_7$$

q<10>=(b<0>-b<2>-b<4>+b<5>-b<6>+b<7>)

$$Q_{10} = B_0 - B_2 - B_4 + B_5 - B_6 + B_7$$

n<10>=p<10>*q<10>

$$N_{10} = P_{10} \, Q_{10}$$

p<11>=(a<0>-a<1>-a<3>+a<5>+a<7>-a<8>)

$$P_{11} = A_0 - A_1 - A_3 + A_5 + A_7 - A_8$$

q<11>=(b<0>-b<1>-b<3>+b<5>+b<7>-b<8>)

$$Q_{11} = B_0 - B_1 - B_3 + B_5 + B_7 - B_8$$

Use of SCRATCHPAD in Algorithms 177

ASILBGR CONSOLE A0 dated 84/04/14 16:35:50 ..... Page 4

n<11>=p<11>*q<11>

$$N_{11} = P_{11} Q_{11}$$

p<12>=(a<1>-a<2>+a<3>-a<4>-a<6>+a<8>)

$$P_{12} = A_1 - A_2 + A_3 - A_4 - A_6 + A_8$$

q<12>=(b<1>-b<2>+b<3>-b<4>-b<6>+b<8>)

$$Q_{12} = B_1 - B_2 + B_3 - B_4 - B_6 + B_8$$

n<12>=p<12>*q<12>

$$N_{12} = P_{12} Q_{12}$$

p<13>=(a<0>-a<1>+a<4>-a<5>-a<6>+a<8>)

$$P_{13} = A_0 - A_1 + A_4 - A_5 - A_6 + A_8$$

q<13>=(b<0>-b<1>+b<4>-b<5>-b<6>+b<8>)

$$Q_{13} = B_0 - B_1 + B_4 - B_5 - B_6 + B_8$$

n<13>=p<13>*q<13>

$$N_{13} = P_{13} Q_{13}$$

p<14>=(a<0>-a<2>-a<3>+a<4>-a<7>+a<8>)

$$P_{14} = A_0 - A_2 - A_3 + A_4 - A_7 + A_8$$

q<14>=(b<0>-b<2>-b<3>+b<4>-b<7>+b<8>)

$$Q_{14} = B_0 - B_2 - B_3 + B_4 - B_7 + B_8$$

n<14>=p<14>*q<14>

$$N_{14} = P_{14} Q_{14}$$

p<15>=(-a<1>+a<2>+a<3>-a<5>-a<6>+a<7>)

$$P_{15} = -A_1 + A_2 + A_3 - A_5 - A_6 + A_7$$

q<15>=(-b<1>+b<2>+b<3>-b<5>-b<6>+b<7>)

$$Q = -B + B + B - B - B + B$$

ASILBGR CONSOLE A0 dated 84/04/14 16:35:50 ..... Page 5

```
        15    1    2    3    5    6    7
n<15>=p<15>*q<15>
```

$$N_{15} = P_{15} Q_{15}$$

p<16>=(a<2>-a<8>)

$$P_{16} = A_2 - A_8$$

q<16>=(b<2>-b<8>)

$$Q_{16} = B_2 - B_8$$

n<16>=p<16>*q<16>

$$N_{16} = P_{16} Q_{16}$$

p<17>=(a<2>-a<5>)

$$P_{17} = A_2 - A_5$$

q<17>=(b<2>-b<5>)

$$Q_{17} = B_2 - B_5$$

n<17>=p<17>*q<17>

$$N_{17} = P_{17} Q_{17}$$

p<18>=(a<5>-a<8>)

$$P_{18} = A_5 - A_8$$

q<18>=(b<5>-b<8>)

$$Q_{18} = B_5 - B_8$$

n<18>=p<18>*q<18>

$$N_{18} = P_{18} Q_{18}$$

"check that the algorithm is correct."
h.0

$$(14) \quad 9B_8 A_1 + 9B_7 A_2 + 9B_6 A_3 + 9B_5 A_4 + 9B_4 A_5 + 9B_3 A_6 + 9B_2 A_7 + 9B_1 A_8$$

Use of SCRATCHPAD in Algorithms 179

ASILBGR CONSOLE A0 dated 84/04/14 16:35:50 ..... Page 6


```
         +
           9B A
            0 0
h.1
         (15)
           9B A   + 9B A   + 9B A   + 9B A   + 9B A   + 9B A   + 9B A   + 9B A
            8 2     7 3     6 4     5 5     4 6     3 7     2 8     1 0
         +
           9B A
            0 1
h.2
         (16)
           9B A   + 9B A   + 9B A   + 9B A   + 9B A   + 9B A   + 9B A   + 9B A
            8 3     7 4     6 5     5 6     4 7     3 8     2 0     1 1
         +
           9B A
            0 2
h.3
         (17)
           9B A   + 9B A   + 9B A   + 9B A   + 9B A   + 9B A   + 9B A   + 9B A
            8 4     7 5     6 6     5 7     4 8     3 0     2 1     1 2
         +
           9B A
            0 3
h.4
         (18)
           9B A   + 9B A   + 9B A   + 9B A   + 9B A   + 9B A   + 9B A   + 9B A
            8 5     7 6     6 7     5 8     4 0     3 1     2 2     1 3
         +
           9B A
            0 4
h.5
         (19)
           9B A   + 9B A   + 9B A   + 9B A   + 9B A   + 9B A   + 9B A   + 9B A
            8 6     7 7     6 8     5 0     4 1     3 2     2 3     1 4
         +
           9B A
            0 5
h.6
         (20)
           9B A   + 9B A   + 9B A   + 9B A   + 9B A   + 9B A   + 9B A   + 9B A
```

ASILBGR CONSOLE A0 dated 84/04/14 16:35:50 ..... Page 7

$$9B A + 9B A + 9B A + 9B A + 9B A + 9B A + 9B A$$
$$8\,7 \quad7\,8 \quad6\,0 \quad5\,1 \quad4\,2 \quad3\,3 \quad2\,4 \quad1\,5$$
$$+\ 9B A$$
$$0\,6$$

h.7

(21)
$$9B A + 9B A + 9B A + 9B A + 9B A + 9B A + 9B A + 9B A$$
$$8\,8 \quad7\,0 \quad6\,1 \quad5\,2 \quad4\,3 \quad3\,4 \quad2\,5 \quad1\,6$$
$$+\ 9B A$$
$$0\,7$$

h.8

(22)
$$9B A + 9B A + 9B A + 9B A + 9B A + 9B A + 9B A + 9B A$$
$$8\,0 \quad7\,1 \quad6\,2 \quad5\,3 \quad4\,4 \quad3\,5 \quad2\,6 \quad1\,7$$
$$+\ 9B A$$
$$0\,8$$

p=(p<i> for i in (0,...,18))

  P= (P   FOR I IN (0,...,18))
      I

a=(a<j> for j in (0,...,8))

  A= (A   FOR J IN (0,...,8))
      J

"this is the matrix b."
coeff(a,p)

```
            *1  1  1  1  1  1  1  1*
            *                      *
            *1  0 -1  1  0 -1  1  0 -1*
            *                      *
            *1 -1  0  1 -1  0  1 -1  0*
            *                      *
            *0  1 -1  0  1 -1  0  1 -1*
            *                      *
            *1  0  0  0  0  0 -1  0  0*
            *                      *
            *1  0  0 -1  0  0  0  0  0*
            *                      *
            *0  0  0  1  0  0 -1  0  0*
            *                      *
            *1  1  1  0  0  0 -1 -1 -1*
            *                      *
            *1  1  1 -1 -1 -1  0  0  0*
            *                      *
     (23)   *0  0  0  1  1  1 -1 -1 -1*
```

ASILBGR CONSOLE A0 dated 84/04/14 16:35:50 ..... Page 8

```
     *                             *
     *1  0  -1  0  -1  1  -1  1   0 *
     *                             *
     *1 -1  0  -1  0  1   0  1  -1 *
     *                             *
     *0  1  -1  1  -1  0  -1  0   1 *
     *                             *
     *1 -1  0   0  1  -1 -1  0   1 *
     *                             *
     *1  0  -1  -1  1  0   0  -1  1 *
     *                             *
     *0 -1  1   1  0  -1 -1  1   0 *
     *                             *
     *0  0  1   0  0   0  0  0  -1 *
     *                             *
     *0  0  1   0  0  -1  0  0   0 *
     *                             *
     *0  0  0   0  0   1  0  0  -1 *
```

ENDOFFILE : RAT9

REFERENCES

1. R. C. Agarwal and J. W. Cooley, New algorithms for digital convolution, IEEE Trans. Acoust. Speech Signal Process. *ASSP-25*(5) (1977).

2. S. Winograd, *Arithmetic Complexity of Computations* (CBMS-NSF Conference Series in Applied Mathematics 33), SIAM, Philadelphia, 1980.

Automated Generation of Optimized Convolution Algorithms

JAMES W. COOLEY   IBM Thomas J. Watson Research Center, Yorktown
Heights, New York

Winograd's theory of computational complexity has provided a theoretical foundation and constructive methods for designing optimized computational algorithms for a very important class of problems. An essential tool in the implementation of these methods is a computer-based formula manipulation system. A brief description of the fields of applications, including the calculation of convolutions and Fourier transforms, is described with a few of the main theorems which are used. An example is worked out in some detail. Programs for using the output of a formula manipulation system to generate, automatically, efficient computer programs are described. These programs accept a set of expressions to be evaluated, seek common subexpressions, and sequence them in an optimal ordering for computational efficiency. It is suggested that this type of formula manipulation could be handled in more generality by implementing some new functions in a formula-processing system.

1. INTRODUCTION

The computational complexity theory developed by Winograd [10-12] applies abstract algebra to the development of a theory of computing. A model of computing is assumed and algebraic methods are used to obtain important theorems concerning the minimization of the number of multiplications required to perform certain computational tasks and the form that optimal algorithms must take. Perhaps even more important, the theory provides algebraic methods for designing optimized algorithms for computing trilinear forms, convolutions, and Fourier transforms. All of these find important applications in digital signal processing and other computation-intensive tasks, where typically, the computer programs must run at very high speed with very large amounts of data, making it worthwhile to expend considerable effort in algorithm design and program development.

The main effort in designing the algorithms discussed here is in carrying out algebraic operations in rings of polynomials modulo polynomials. While doing so, one uses not only theory but insight gained by observing the results of the manipulations while trying many possible choices of parameters and strategies. Some illustrative examples for the calculation of short convolutions can easily be carried out with pencil and paper, but as soon as one attempts to work with convolutions of useful sizes, one is severely limited by the time and labor required to carry out the polynomial manipulations. For problems of even modest size, an automatic formula manipulation system is essential. It must, of course, be interactive since an essential feature of the calculation is that one wants to observe the effects of various strategies while planning subsequent steps. At times, an entire procedure can be developed in this manner and the resulting program can be made to do the whole task over again in a "batch" mode with slight changes in the program in order to try different alternatives. All of the desired features noted above were found in the SCRATCHPAD formula manipulation system at the IBM Watson Research Center [6].

The SCRATCHPAD sessions were designed so that their final outputs were matrices that were in disk storage. These became inputs to other programs written by the author, which then optimized the sequencing of the additions and produced, on disk, PLI programs for implementing the algorithms. These programs were then analyzed automatically for efficiency and error generation. The entire procedure provided an interactive system for designing and producing optimized programs in which strategy and decision making could be performed at the computer terminal, with all the tedious and laborious parts of the effort being carried out by the computer. It may be practical, in the future, to develop this type of program into a system that produces programs in a language suitable for input to a VLSI compiler such as that developed by Brayton et al. [5]. This system optimizes a VLSI design for carrying out the calculation and produces output that can be used to produce a VLSI chip automatically.

Further details on the use of SCRATCHPAD in the derivation of convolution algorithms are given in the chapter by Auslander and Silberger [1].

2. THE USES OF SCRATCHPAD-GENERATED CONVOLUTION ALGORITHMS

The algorithms designed as described above are usually modules that are used in programs performing two major classes of computations which, since they are described elsewhere, will be treated only briefly.

Convolutions

It has been shown by Agarwal and Burrus [2,3] and by Agarwal and Cooley [4] how one can compute a long convolution by several different techniques which essentially map it into a multidimensional convolution. The dimensions are the factors of the length of the long convolution. A program would consist of modules that compute convolutions whose lengths are these factors. In this manner, the "divide and conquer" approach is to compute the long convolutions via the relatively short convolutions. The latter are developed and optimized by the SCRATCHPAD sessions and incorporated into the full program as modules.

Fourier Transforms

The basic theorems of Winograd's computational complexity theory deal with trilinear forms, of which the convolution is an important special case. In addition, Winograd has shown how one can divide and conquer a large Fourier transform calculation by special permutations of the indices which convert the Fourier transform matrix into a matrix of block matrices each of which corresponds to a convolution of the data with a set of constants obtained from the elements of the Fourier transform matrix. Each of these convolutions can then be computed by a program obtained from the SCRATCHPAD derivations described below.

There have been two general methods for modularizing the Fourier transform:

1. The nesting method as designed by Winograd [11] and programmed by Silverman [9]. This reduces the number of multiplications by combining all of them in one innermost loop.
2. The methods of Kolba and Parks [7], which result in a more modular construction of the program. It turns out that, due to programming considerations, the latter has found preference, although this will not necessarily be true for all program environments.

Although the convolutions usually computed are not cyclic, computational economy makes it useful to section data and compute cyclic convolutions of overlapping segments or segments extended with zeros. The various divide-and-conquer approaches then lead to the need for smaller cyclic convolutions of the sizes produced by the methods described below.

3. MATHEMATICAL BACKGROUND

We define a convolution of two sequences, $h_0, \ldots, h_{N-1}$ and $x_0, \ldots, x_{N-1}$ as the sequence

$$y_j = \sum_{i=0}^{N-1} h_i x_{j-i} \tag{3.1}$$

$i = 0, \ldots, N - 1$. If the sequences h_i and x_i are periodic, of period N, the convolution is *cyclic*; if not, it is called *noncyclic*. In the theory, and in what follows, convolution is described in terms of polynomial multiplication by letting

$$H(u) = \sum_{i=0}^{N-1} h_i u^i \tag{3.2}$$

with similar definitions for $X(u)$ and $Y(u)$. Thus the sequence y_j is given by the coefficients of

$$Y(u) = H(u)X(u) \tag{3.3}$$

(This is essentially the z-transform method. It differs from electrical engineering notation in that here positive rather than negative powers of z are used.) We consider operations in a ring $Q[u]$ of polynomials in u with coefficients in the field Q of rational numbers (i.e., the field containing the data). We count, as a multiplication, a multiplication of two numbers in the data field. We also consider rings $Q[u]/(P(u))$ defined as residue classes of polynomials modulo polynomials $P(u)$ in $Q[u]$. By an *irreducible polynomial* we mean a polynomial with coefficients in Q which cannot be expressed as a product of polynomials in $Q[u]$.

In particular, a *cyclic convolution* of period N is equivalent to a polynomial product,

$$Y(u) = H(u)X(u) \mod(u^N - 1) \tag{3.4}$$

in $Q[u]/(u^N - 1)$. This is equivalent to regarding the sequences in (3.1) as periodic of period N.

Several of Winograd's theorems are prominent in the divide-and-conquer approach. First, we have:

THEOREM 3.1 If $P(u)$ is an irreducible polynomial of degree N, then the minimum number of multiplications required to compute

$$Y(u) = H(u)X(u) \mod P(u) \tag{3.5}$$

is $2N - 1$.

If

$$P(u) = P_1(u), \ldots, P_K(u) \tag{3.6}$$

where the $P_i(u)$'s are irreducible, the Chinese remainder theorem enables one to prove that:

THEOREM 3.2 If $P(u)$ has K factors, then the minimum number of multiplications required to compute (3.5) is $2N - K$.

If $P(u) = u^N - 1$ and there are K factors of N, including 1 and N, the irreducible factors of $u^N - 1$ are the "cyclotomic polynomials." These form a special set of polynomials which have been studied by number theorists (Ref. 8) and are shown to have coefficients which are 0's and ±1's for polynomials up to a very high degree. In this case, we have:

THEOREM 3.3 The minimum number of multiplications required to compute

$$Y(u) = H(u)X(u) \mod(u^N - 1) \tag{3.7}$$

is $2N - K$, where K is the number of factors of N, including 1 and N.

Finally, letting $\underline{h}$, $\underline{x}$, and $\underline{y}$ denote the sequences in (3.1), we express the convolution in vector notation

$$\underline{y} = \underline{h} * \underline{x} \tag{3.8}$$

A theorem of Winograd then states:

THEOREM 3.4 The algorithm for computing the polynomial product (3.7) or, equivalently, the cyclic convolution (3.1) in the minimum number of multiplications M has the form

$$\underline{y} = \underline{C}(\underline{A}\underline{h}) \times (\underline{B}\underline{x}) \tag{3.9}$$

where $\underline{A}$ and $\underline{B}$ are M by N matrices and $\underline{C}$ is an N by M matrix. The "×" denotes element-by-element multiplication of the vectors. The resemblance to Fourier transform methods for computing convolutions, with the rectangular matrices taking the place of the square Fourier transform matrix, has led to the term "rectangular transform" (RT) for these techniques. It turns out that, in general, one does not find the optimal algorithm but instead makes some trade-offs to obtain other favorable properties of the final program. In this case, we say that we have a *suboptimal* algorithm. It turns out that all suboptimal algorithms can also be put in the form (3.9). The final important point to be made here is that the final output of the SCRATCHPAD sessions consists of the matrices $\underline{A}$, $\underline{B}$, and $\underline{C}$.

4. DERIVATION OF THE N = 7 ALGORITHM

We demonstrate the mathematical derivation of an N = 7 RT convolution algorithm. Those who are familiar with any formula manipulation system will easily see how the procedure may be implemented. As described above, we first factor

$$P(u) = u^7 - 1 = P_1(u) P_2(u) \tag{4.1}$$

giving

$$P_1(u) = u^6 + u^5 + u^4 + u^3 + u^2 + u + 1 \tag{4.2}$$

$$P_2(u) = u - 1 \tag{4.3}$$

In general, superscripts will denote polynomial residues, as in

$$X^j(u) = \sum_i x_i^j u^i = X(u) \mod P_j(u) \tag{4.4}$$

This gives

$$X^1(u) = \sum_{i=0}^{5} x_i^1 u^i \quad \text{where } x_i^1 = x_i - x_6 \tag{4.5}$$

and

$$X^2(u) = x_0^2 = \sum_{i=0}^{6} x_i \tag{4.6}$$

The Chinese remainder theorem (CRT) says that

$$Y(u) = Y^1(u) e_1(u) + Y^2(u) e_2(u) \mod(u^7 - 1) \tag{4.7}$$

where

$$Y^j(u) = H^j(u) X^j(u) \mod P_j(u) \quad j = 1, 2 \tag{4.8}$$

and where the $e_j(u)$'s are idempotents of $Q[u]/(u^7 - 1)$ and satisfy

$$1 = e_1(u) + e_2(u) \mod(u^7 - 1) \tag{4.9}$$

$$e_1(u) = 1 \mod P_1(u) \qquad e_1(u) = 0 \mod P_2(u)$$

$$e_2(u) = 1 \mod P_2(u) \qquad e_2(u) = 0 \mod P_1(u)$$

From these, one can derive

$$e_1(u) = -\frac{(u-1)(u^5 + 2u^4 + 3u^3 + 4u^2 + 5u + 6)}{7} \tag{4.10}$$

$$e_2(u) = \frac{u^6 + u^5 + u^4 + u^3 + u^2 + u + 1}{7} \tag{4.11}$$

In SCRATCHPAD, it is not necessary to be concerned with the idempotents when applying the CRT. David Yun of the SCRATCHPAD project set up a polynomial Chinese remainder algorithm user command (PCRA) which allows one to obtain the result directly by writing a command which is a SCRATCHPAD equivalent of

$$Y(u) = PCRA(Y^1(u), Y^2(u); P_1(u), P_2(u)) \tag{4.12}$$

(One can have more than two arguments in the PCRA command.)

The polynomial

$$\tilde{Y}(u) = H^1(u) X^1(u) \tag{4.13}$$

which is needed in obtaining $Y^1(u)$, is a product of two-fifth-degree polynomials and is therefore a tenth-degree polynomials which is equivalent to a noncyclic six-point convolution. If we let $F(u)$ be any polynomial of degree 11, then

$$\tilde{Y}(u) = \tilde{Y}(u) \mod F(u) \tag{4.14}$$

and one can compute the formula for $\tilde{Y}(u)$ by choosing an $F(u)$ with many simple factors, reducing the polynomials mod the factors, performing the small polynomial multiplications, and then putting the results together by means of the CRT. Then all one has to do is generate the remainder

$$Y^1(u) = \tilde{Y}(u) \mod P_1(u) \tag{4.15}$$

Several variations on this procedure are possible. One that gives a simpler algorithm is obtained by letting $F(u)$ be of degree 9. Then one can express $\tilde{Y}(u)$ in terms of its quotient,

$$Q(u) = \tilde{y}_9 + \tilde{y}_{10}u \tag{4.16}$$

and remainder, when divided by $F(u)$, as defined by the expression

$$\tilde{Y}(u) = Q(u)F(u) + \tilde{Y}(u) \mod F(u) \tag{4.17}$$

The remainder, in the second term of (4.17), may be obtained by the CRT.

The polynomial $F(u)$ is chosen so as to make the resulting algorithm as simple as possible. In general, it may require some insight and experience in choosing $F(u)$ under the general constraints mentioned above. One of the choices considered and used as an example here is

$$\begin{aligned} F(u) &= (u-1)(u+1)u(u^2+u+1)(u^2-u+1)(u^2+1) \\ &= u^9 + u^7 - u^3 - u \end{aligned} \tag{4.18}$$

It should be mentioned that this did not turn out to be the best choice; the root $u = 1$ produced large coefficients in the $\underline{A}$ and $\underline{B}$ matrices. It can easily be seen that this will always happen, so the root $u = 1$ was avoided in subsequent derivations. Letting $F_v(u)$, $v = 1, 2, \ldots, 6$ denote the factors in (4.18), we use second superscripts on the polynomial residues

$$X^{j,k}(u) = \Sigma_i \, x_i^{j,k} u^i = X^j(u) \mod F_k(u) \tag{4.19}$$

with similar definitions of $H^{j,k}(u)$ in terms of h's. Letting

$$\tilde{Y}^k(u) = \tilde{Y}(u) \mod F_k(u) \tag{4.20}$$

the CRT gives the remainder term in the noncyclic convolution (4.17) by means of the SCRATCHPAD command

$$\tilde{Y}(u) \mod F(u) = \text{PCRA}(\tilde{Y}^1(u), \ldots, \tilde{Y}^6(u); F_1(u), \ldots, F_6(u)) \tag{4.21}$$

For this, we must now derive expressions for the $\tilde{Y}^k(u)$'s in terms of the h_i's and x_i's.

The coefficients of $X^{1k}(u)$ are defined in terms of the coefficients of $X^1(u)$ by (4.19), giving

$$\begin{aligned} x_0^{1,1} &= x_0^1 + x_1^1 + x_2^1 + x_3^1 + x_4^1 + x_5^1 \\ x_0^{1,2} &= x_0^1 - x_1^1 + x_2^1 - x_3^1 + x_4^1 - x_5^1 \\ x_0^{1,3} &= x_0^1 \end{aligned} \tag{4.22}$$

$$x_0^{1,4} = x_0^1 - x_2^1 + x_3^1 - x_5^1$$

$$x_1^{1,4} = x_1^1 - x_2^1 + x_3^1 - x_5^1$$

$$x_0^{1,5} = x_0^1 - x_2^1 - x_3^1 + x_5^1 \qquad (4.22)$$
$$\text{cont.}$$

$$x_1^{1,5} = x_1^1 + x_2^1 - x_3^1 - x_5^1$$

$$x_0^{1,6} = x_0^1 - x_2^1 + x_4^1$$

$$x_1^{1,6} = x_1^1 - x_3^1 + x_5^1$$

The derivation of these formulas by SCRATCHPAD or any other formula manipulation system is fairly straightforward. The $h_i^{j,k}$'s are obtained by replacing all x's by h's. In terms of the quantities above, we obtain expressions for the polynomial residues:

$$\tilde{Y}^k(u) = \tilde{Y}(u) \mod F_k(u) \qquad (4.23)$$

$$\tilde{Y}^1(u) = h_0^{1,1} x_0^{1,1}$$

$$\tilde{Y}^2(u) = h_0^{1,2} x_0^{1,2}$$

$$\tilde{Y}^3(u) = h_0^{1,3} x_0^{1,3} \qquad (4.24)$$

$$\tilde{Y}^4(u) = h_0^{1,4} x_0^{1,4} - h_1^{1,4} x_1^{1,4} + (h_0^{1,4} x_1^{1,4} + h_1^{1,4} x_0^{1,4} - h_1^{1,4} x_1^{1,4})u$$

$$\tilde{Y}^5(u) = h_0^{1,5} x_0^{1,5} - h_1^{1,5} x_1^{1,5} + (h_0^{1,5} x_1^{1,5} + h_1^{1,5} x_0^{1,5} + h_1^{1,5} x_1^{1,5})u$$

$$\tilde{Y}^6(u) = h_0^{1,6} x_0^{1,6} - h_1^{1,6} x_1^{1,6} + (h_0^{1,6} x_1^{1,6} + h_1^{1,6} x_0^{1,6})u$$

Expressions for the coefficients of the quotient in (4.16) are simply

$$\tilde{y}_9 = h_4^1 x_5^1 + h_5^1 x_4^1 \qquad (4.25)$$

$$\tilde{y}_{10} = h_5^1 x_5^1 \qquad (4.26)$$

Finally, there is only one multiplication in the calculation of

$$Y^2(u) = H^2(u)X^2(u) = (h_0 + h_1 + \cdots + h_6)(x_0 + x_1 + \cdots + h_6) \qquad (4.27)$$

One may note that the calculation of $\tilde{Y}^6(u)$ is equivalent to complex multiplication, with u playing the role of i. For this we can use any of the well-known algorithms for complex multiplication in three instead of four real multiplications. Each of the products $\tilde{Y}^4(u)$ and $\tilde{Y}^5(u)$ can also be computed in three multiplications. Thus the following 16 products are required:

$$
\begin{aligned}
m_0 &= h_0^{1,1} x_0^{1,1} \\
m_1 &= h_0^{1,2} x_0^{1,2} \\
m_2 &= h_0^{1,3} x_0^{1,3} \\
m_3 &= h_0^{1,4} x_0^{1,4} \\
m_4 &= h_1^{1,4} x_1^{1,4} \\
m_5 &= (h_0^{1,4} - h_1^{1,4})(x_0^{1,4} - x_1^{1,4}) \\
m_6 &= h_0^{1,5} x_0^{1,5} \\
m_7 &= h_1^{1,5} x_1^{1,5} \\
m_8 &= (h_0^{1,5} + h_1^{1,5})(x_0^{1,5} + x_1^{1,5}) \\
m_9 &= h_0^{1,6} x_0^{1,6} \\
m_{10} &= h_1^{1,6} x_1^{1,6} \\
m_{11} &= (h_0^{1,6} + h_1^{1,6})(x_0^{1,6} + x_1^{1,6}) \\
m_{12} &= h_4^1 x_5^1 \\
m_{13} &= h_5^1 x_4^1 \\
m_{14} &= h_5^1 x_5^1 \\
m_{15} &= h_0^2 x_0^2
\end{aligned}
\tag{4.28}
$$

In terms of these products, (4.24), (4.25), and (4.26) can be written

$$
\begin{aligned}
\tilde{Y}^1(u) &= m_0 \\
\tilde{Y}^2(u) &= m_1 \\
\tilde{Y}^3(u) &= m_2 \\
\tilde{Y}^{1,4}(u) &= m_3 - m_4 + (m_3 - m_5)u
\end{aligned}
\tag{4.29}
$$

$$\tilde{Y}^{1,5}(u) = m_6 - m_7 + (m_8 - m_6)u$$

$$\tilde{Y}^{1,6}(u) = m_9 - m_{10} + (m_{11} - m_9 - m_{10})u$$

$$\tilde{Y}^1_9 = Y^1_{10}u = m_{12} + m_{13} + m_{14}u$$

$$Y^2(u) = m_{15}$$

(4.29) cont.

Each m_n in (4.28) is a product of a linear combination of h_j's and a linear combination of x_i's. In the notation, these will be denoted by a_n's and b_n's, respectively, so that

$$m_n = a_n b_n \qquad \text{for } n = 0, 1, \ldots, 15 \tag{4.30}$$

where

$$
\begin{aligned}
a_0 &= h_0^{1,1} & b_0 &= x_0^{1,1} \\
a_1 &= h_0^{1,2} & b_1 &= x_0^{1,2} \\
a_2 &= h_0^{1,3} & b_2 &= x_0^{1,3} \\
a_3 &= h_0^{1,4} & b_3 &= x_0^{1,4} \\
a_4 &= h_1^{1,4} & b_4 &= x_1^{1,4} \\
a_5 &= a_3 - a_4 & b_5 &= b_3 - b_4 \\
a_6 &= h_0^{1,5} & b_6 &= x_0^{1,5} \\
a_7 &= h_1^{1,5} & b_7 &= x_1^{1,5} \\
a_8 &= a_6 + a_7 & b_8 &= b_6 + b_7 \\
a_9 &= h_0^{1,6} & b_9 &= x_0^{1,6} \\
a_{10} &= h_1^{1,6} & b_{10} &= x_1^{1,6} \\
a_{11} &= a_9 + a_{10} & b_{11} &= b_9 + b_{10} \\
a_{12} &= h_4^1 & b_{12} &= x_5^1 \\
a_{13} &= h_5^1 & b_{13} &= x_4^1 \\
a_{14} &= h_5^1 = a_{13} & b_{14} &= x_5^1 = b_{12} \\
a_{15} &= h_0^2 & b_{15} &= x_0^2
\end{aligned}
$$

(4.31)

In vector-matrix notation, (4.31) and (4.30) can be written

$$\underline{a} = \underline{A}\underline{h} \qquad \underline{b} = \underline{B}\underline{x} \qquad (4.32)$$

$$\underline{m} = \underline{a} \times \underline{b} \qquad (4.33)$$

respectively.

Several properties of $\underline{A}$ and $\underline{B}$ should be noted for future reference. One is that $\underline{A}$ and $\underline{B}$ are the same except for the interchange of rows 12 and 13. Another is that rows 5, 8, 11, and 14 are each given in terms of previous rows.

The procedure for obtaining the a_n's and the b_n's in (4.31) is apparent. One merely has to define the polynomials $H(u)$ and $X(u)$ with their coefficients as indeterminants. Then one finds the remainders of $H(u)$ and $X(u)$ with respect to the various factors of $u^7 - 1$ and $F(u)$. From these, one obtains the a_n's and b_n's, respectively, in (4.31) as linear combinations of the h_i's and x_i's, respectively. The matrices of coefficients are the $\underline{A}$ and $\underline{B}$ matrices of (3.9), which are given as final outputs of the SCRATCHPAD session. Actually, one only computes the $\underline{B}$ matrix. Interchanging rows 12 and 13 in $\underline{B}$ gives the matrix $\underline{A}$.

The procedure for obtaining the $\underline{C}$ matrix, giving the convolution $\underline{y}$ in terms of the m_n's may be summarized as follows: The polynomial residues are expressed in terms of the m_n's according to (4.29). The CRT is used as in (4.21) to obtain $\tilde{Y}(u) \bmod F(u)$, which may be substituted in (4.17) to give an expression for $\tilde{Y}(u)$ in terms of the m_n's. From this, one gets

$$Y^1(u) = \tilde{Y}(u) \bmod P_1(u) \qquad (4.34)$$

This, in turn, is used with the CRT in (4.12) to express $Y(u)$ as a polynomial in u whose coefficients are a set of expressions for the values of the circular convolution in terms of the m_n's. By their construction, it is seen that this is a linear dependence and that the final result can be written

$$\underline{y} = \underline{C}\underline{m} \qquad (4.35)$$

where $\underline{C}$ is a 7×16 matrix.

5. POST-SCRATCHPAD PROCESSING

We consider the case where the $\underline{h}$ sequence is constant and is convolved with many $\underline{x}$ sequences which are usually sections of data from a long digitized analog signal. These are overlapped or padded with zeros to permit the use of cyclic convolution algorithms. Therefore, a theorem of Winograd [12] is used which permits one to exchange the simpler $\underline{A}$ matrix for the complicated $\underline{C}$ matrix and express the algorithm in the form

$$\underline{y} = \underline{A}'(\underline{C}'\underline{h}) \times (Bx) \tag{5.1}$$

where $\underline{A}$ and $\underline{C}$ represent special types of transposes of $\underline{A}$ and $\underline{C}$ as described in Ref. 4 and in [12]. In this way, $\underline{C}'\underline{h}$ is computed once and each $\underline{x}$ segnent is processed by computing the $\underline{B}$ transform, performing the multiplication in (5.1), and then computing the $\underline{A}'$ transform of the result.

The post-SCRATCHPAD processing program performs this transposition. It then searches for and counts common subexpressions, represents them by new variables, and proceeds to nest the additions in such a manner as to minimize the number of additions. A part of the strategy uses information about linear dependencies of rows and columns of $\underline{A}$ and $\underline{B}$. Certain groupings of variables formed by the Chinese remainder theorem decomposition is sometimes useful. The program also uses heuristics and techniques developed while doing a number of examples manually. Unfortunately, no theorems or theory was found which minimized additions as was the case for multiplications. However, the programs always did at least as well or better than the manual methods. However, they did the task much faster, so that one could observe the results and return to the SCRATCHPAD program to try variations on the procedure.

6. A SUGGESTED EXTENSION OF SCRATCHPAD

The post-SCRATCHPAD processing does the inverse of what SCRATCHPAD does very well. Whereas SCRATCHPAD substitutes subexpressions to form larger expressions, we start with a set of large expressions and try to define common subexpressions so that the resulting program will run as fast as possible. An added requirement could be that the subexpressions be defined so that as many as possible can be computed concurrently on a parallel processor of some kind. The present program involved only additions and turned out to be very useful. It is suggested here that a very useful facility in a formula manipulations system would be to have it do, in a more general way, the postprocessing described here and give, as final output, computer programs that are optimized for some given type of computer.

7. CONCLUSIONS

Winograd's computational complexity theory has given some theorems and constructive methods for optimizing an important class of computational algorithms. It may very well be that this is only a beginning which may show the way to even more important theoretical results. However, very soon in the implementation of this theory, one becomes limited by the tedious algebraic manipulations that are required to do the derivations. Therefore, to make these and future theoretical results useful, it is absolutely essential to be able to perform these formula manipulations automatically on an interactive system. Furthermore, the results of a complicated formula manipulation system are seldom expected to terminate in a set of formulas that one simply views on the screen or prints on paper. They will often have to be implemented in a computer program which, hopefully, will be efficient enough to make the whole effort worthwhile.

Since a formula manipulations system already has much of the machinery necessary to manipulate expressions, it seems reasonable to expect future extensions to perform optimized expression evaluation and generate efficient operable programs as a final product. Such a system could help overcome some of the objections to the new and "faster" algorithms, where it has been argued at conferences and in the literature that they require costly program development and result in complicated programs and costly overhead.

REFERENCES

1. L. Auslander and A. Silberger, On the Use of SCRATCHPAD in the Construction of Convolution Algorithms, Research Report RC 10554, IBM Watson Research Center, Yorktown Hts., New York, May 1984.

2. R. C. Agarwal and C. S. Burrus, Fast convolution using Fermat number transforms with applications to digital filtering, IEEE Trans. Acoust. Speech Signal Process. *ASSP-22*(2) (1974), 87-99. Also in Proc. IEEE *63* (1975), 550-560.

3. R. C. Agarwal and C. S. Burrus, Fast one-dimensional digital convolution by multidimensional techniques, IEEE Trans. Acoust. Speech Signal Process. *ASSP-25*(1) (1974), 1-10.

4. R. C. Agarwal and J. W. Cooley, New algorithms for digital convolution, IEEE Trans. Acoust. Speech, Signal Process. *ASSP-25*(5) (1977), 392-410.

5. R. K. Brayton, C. L. Chen, C. T. McMullen, R. H. J. M. Otten, and Y. J. Yamour, *Automated Implementation of Switch Functions as Dynamic CMOS Circuits*, Proceedings of the 1984 Custom Integrated Circuit Conference, pp. 346-350.

6. J. H. Griesmer, R. D. Jenks, and D. Y. Y. Yun, *SCRATCHPAD User's Manual*, Research Report RA70, IBM Watson Research Center, Yorktown Heights, N.Y., June 1975.

7. D. P. Kolba and T. W. Parks, A prime factor FFT algorithm using high-speed convolution, IEEE Trans. Acoust. Speech Signal Process. *ASSP-25* (4) (1977), 281-294.

8. T. Nagelle, *Introduction to Number Theory*, John Wiley, New York, 1951.

9. H. F. Silverman, An introduction to programming the Winograd Fourier transform algorithm (WFTA), IEEE Trans. Acoust. Speech Signal Process. *ASSP-25*(2) (1977), 152-164.

10. S. Winograd, *Some Bilinear Forms Whose Multiplicative Complexity Depends on the Field of Constants*, Research Report RC5669, IBM Watson Research Center, Yorktown Heights, N.Y., Oct. 10, 1975.

11. S. Winograd, On computing the discrete Fourier transform, Proc. Nat. Acad. Sci. USA *73*(4) (1976), 1005-1006. Also in Math. Comp. *32*(141) (1978), 175-199.

12. S. Winograd, *Arithmetic Complexity of Computations* (CBMS-NSF Conference Series in Applied Mathematics 33), SIAM, Philadelphia, 1980.

Manual for the System PNCRE

ROBERT RILEY State University of New York at Binghamton, Binghamton, New York

The P (= PNCRE) system is a file of 44 Fortran subroutines which are to be called by a main program. This main program would somehow set up a list $\underline{X} = \{X_1,\ldots,X_m\}$ of matrices in $SL_2(\mathbb{C})$ which generate a group Γ. The purpose of P is to make decisions about Γ, and in favorable cases, to calculate various quantities for Γ. These are

1. A decision whether or not Γ is discrete. If Γ is found to be not discrete, the P system can do nothing more with it, and if Γ is discrete but geometrically infinite, the P system will not complete its work successfully.

When Γ is discrete and geometrically finite P can do the following.

2. Compute a list $\underline{T} = \{T_1,\ldots,T_n\}$ of elements of Γ that generate Γ and act as side pairing transformations of a Ford fundamental domain $\mathcal{D}$ for the action of Γ on a model of hyperbolic space H^3. The domain $\mathcal{D}$ is described by lists of edges, vertices, and sides in a rather indirect way which is convenient for the edge cycle presentation of Γ on $\underline{T}$ corresponding to $\mathcal{D}$. The principal output of P is this list $\underline{T}$ and the edge cycle representation of Γ. The proof that Γ actually is discrete consists of this information plus a proof that the cycle relations actually hold in Γ, and that $\mathcal{D}$ has the correct combinatorial intersection

pattern for the H-planes carrying the sides of $\mathcal{D}$. These last two items are not part of P and will have to be supplied by the user.

3. When $\mathcal{D}$ has been computed using the upper half-space model U of $\underline{H}^3$, P can call a graph plotter to produce a diagram of the normal projection of $\mathcal{D}$ on $\mathbb{C}$, showing the isometric circles and reflecting lines for $\underline{T}$, and the projections of the non EH-edges of $\mathcal{D}$.

4. Let $\pi\Gamma := \Gamma/\Gamma^f$, where Γ^f is the normal subgroup of Γ generated by the elements of Γ which have fixed points in $\underline{H}^3$, that is, $\pi\Gamma = \pi_1(\underline{H}^3/\Gamma;*)$. Then P can produce a presentation for $\pi\Gamma$ on the images of $\underline{T}$ in $\pi\Gamma$ as generators. It tries to get a simple result.

5. When $\Delta \subset SL_2(\mathbb{C})$ is generated by Γ and by EH-transformations, P can get a Ford domain and corresponding edge cycle presentation for Δ by combining the completed work for Γ with the new generators of Δ. This facility is used to get the symmetry groups of hyperbolic space forms of finite volume.

Although the main program calling P sets up the generating set $\underline{X}$ of Γ, P contains subroutines that make this especially convenient in several ways. For example, if the X_ν are naturally expressed as words on the other matrices Y_μ which do not themselves belong to Γ, P has subroutines that do most of the work of setting up the list $\underline{X}$ once $\underline{Y}$ is stored in the system. In fact, a great deal of effort has been spent to make using P as easy as possible, and most of the subroutines in the system are never called directly by the user. One merely has to write a main program to call a certain selection of routines of P according to the pattern described in Section 3, and this selection will invoke the hidden subroutines on your behalf. There are lots of parameters that decide the operating environment of the calculations, and these are given good default values whenever possible. This means that the most standard options for working on Γ do not require knowing about many of these parameters, but P does allow one to use unusual options, at the cost of learning about them.

The system P was conceived at a time when interactive computing was not available, and I have never even considered how one could make interactive computation with P. One major problem is the large bulk of the output of P for a given group Γ. This really needs to be examined in an unhurried manner which only printed output and diagrams can provide.

The P system probably contains mistakes and it almost certainly is unnecessarily inadequate for coping with certain combinations of circumstances that I have not yet seen. It probably contains many inferior choices and segments of code. The outline of the development of P in the context of representations of knot groups given in Section 1 (which also introduces some standard notation and names) will give an idea of what P was originally intended to do and how it evolved from its early versions. One remark must be made: a guaranteed algorithm to find a Ford domain $\mathcal{D}$ for a geometrically finite discrete group Γ described by a list of matrices $\underline{X}$ would act like the mill of the gods, and P was only supposed to get good results for the lists $\underline{X}$ that actually arose in my projects. It will succeed only when $\mathcal{D}$ has limited complexity, where the limits are generous enough for me, but within these limits P is reliable and quite satisfactory.

In Section 2 I describe modifications to the distributed version of the P file which may have to be made by a user during the initial setting up of the system. The available equipment and software for each computer will dictate the exact changes that are necessary, but the kinds of alterations that are likely are mentioned. In Section 3 I give an outline of the features a main program calling P must include, the calls to subroutines of P, some of the parameters that control the operations of P, and a sample segment of code containing the calls to the "heavy" subroutines of P. In Section 4 I comment on a sample main program, KNOTS3, and its attendant subroutines and data file, which are included with the source code for P. It will help debug your implementation of P, and its subroutines constitute an enhancement to P which might have more general uses. The principal reference for background on $SL_2(\mathbb{C})$ and $\underline{H}^3$ for this manual is Riley [3], and the best book on hyperbolic geometry and discrete groups is Beardon [1].

1. HISTORY OF P

In 1974 I realized that it must be true that most knots k in S^3, with torus knots and satellite knots being the only visible exceptions, have the property that the complement $S^3 - k$ can be given the structure of hyperbolic space form of finite volume. This means that the group πK of a typical knot k must have a faithful representation θ in $SL_2(\mathbb{C})$ such that certain elements, the longitudes and meridians, map to parabolics, and the image group $\pi K\theta$ is discrete and geometrically finite. One could check this prediction for a knot k by calculation, and the original purpose of P was to show that $\pi K\theta$ is discrete, that $\underline{H}^3/\pi K\theta$ has finite H-volume, and that $\theta: \pi K \to \pi K\theta$ is faithful.

The earliest version of P was specifically designed to do this for a two-bridge knot. Such a knot is naturally determined by an ordered pair of integers satisfying certain restrictions, and the faithful representation θ is then completely determined by a complex number $\xi = ZR + iZI$. All this survives in the current version of P as the convention that the group Γ being considered by P is designated by an integer pair name (CR,TN) by the output subroutines, and the graph plotter subroutine also requires a real pair ZR, ZI. The connection of these pairs with Γ is now set by the user for each computing project.

Knot groups are most naturally presented with meridians as generators and the representations of greatest interest send meridians to parabolics. One meridian image can conveniently be taken to be

$$A := \begin{bmatrix} 1 & 1 \\ 0 & 1 \end{bmatrix} \quad \text{where } A(\infty) = \infty, \quad A(z) = z + 1 \tag{1.1}$$

Then the image of the longitude for this meridian would be

$$L := \begin{bmatrix} -1 & \lambda \\ 0 & -1 \end{bmatrix} \quad \lambda \in \mathbb{C} \quad \text{so } L(\infty) = \infty, \quad L(z) = z - \lambda$$

But for the action of $\pi K\theta$ on H^3 we may as well consider $\pi K\theta$ as a subgroup of $PSL_2(\mathbb{C})$, in which case

$$L = \begin{bmatrix} 1 & -\lambda \\ 0 & 1 \end{bmatrix}$$

In the current P this survives as the convention that the EH-elements of the group Γ under consideration are of special importance, and to a certain extent are treated differently from the rest of the elements. P now allows Γ to have an EH-translation

$$A = \begin{bmatrix} 1 & A12 \\ 0 & 1 \end{bmatrix} \quad A12 \text{ real}, \quad A12 \text{ positive} \tag{1.2}$$

as its first generator. The default for A is (1.1) (i.e., A12 = 1.0). If a second EH-translation L is present, it is required to be

$$L = \begin{bmatrix} 1 & L12 \\ 0 & 1 \end{bmatrix} \quad L12 \in \mathbb{C} \quad \text{Im}(L12) > 0 \quad |\text{Re}(L12)| \leq \frac{1}{2} A12 \tag{1.3}$$

Although knot group images usually do not contain EH-rotations, the automorphism group $\text{Aut}(\pi K)$ has a faithful representation in $P\Gamma L_2(\mathbb{C})$ when

$S^3 - k$ can be made a hyperbolic space form of finite volume (briefly, when k is *excellent*), and the space-orientation-preserving subgroup $\text{Aut}^+(\pi K\theta)$ often contains an EH-rotation R of order 2. Write

$$R = R\{b\} := \begin{bmatrix} i & b \\ 0 & -i \end{bmatrix} \quad b \in \mathbb{C} \quad \text{so } R(\infty) = \infty, \quad R(z) = -z + ib \quad (1.4)$$

Then the knot group origin of P required P to be able to handle groups Γ whose subgroups Γ_{EH} fixing ∞ are one of

$$\langle A \rangle, \langle A, L \rangle, \langle A, L, R \rangle, \langle A, R \rangle, \langle R \rangle, \langle E \rangle \quad (1.5)$$

But there was no demand for EH-rotations of higher order (see Ref. 4, Sec. 1), and because allowing Γ_{EH} to contain R required a great deal of programming effort (e.g., the long complex subroutine RETEST and an attendant subroutine specifically address complications caused by R), we never got around to allowing for an EH-rotation of order not 2.

The main part of the work done by P for a group Γ is the search for a Ford domain $D \subset H^3$ of Γ. My original guess was that this search was going to be troublesome, and that every assistance that circumstances permit would be needed. (Experience shows that this guess was too pessimistic.) Now two-bridge knot groups have predictable automorphisms, and for an excellent knot these can be realized as conjugations by elements of $P\Gamma L_2(\mathbb{C})$ of the faithful image $\pi K\theta$, according to Mostow's rigidity theorem (see Ref. 2). Accordingly, I had the search make use of conjugations by EH-transformations $A\{h\}$, $R\{b\}$ not belonging to $\pi K\theta$, and also conjugations by glide-reflections J,

$$J(\infty) = \infty \quad J(z) = \varepsilon \bar{z} + \beta \quad \text{where } \varepsilon = \pm 1, \quad \beta \in \mathbb{C}$$

$$J \begin{bmatrix} a & b \\ c & d \end{bmatrix} J^{-1} = \begin{bmatrix} \bar{a} + \varepsilon \beta \bar{c} & \varepsilon(\bar{b} - \bar{c}\beta^2) + \beta(\bar{d} - \bar{a}) \\ \varepsilon \bar{c} & \bar{d} - \varepsilon \beta \bar{c} \end{bmatrix} \quad (1.6)$$

This facility is still part of P. It really does not help the search for D very much, and it causes trouble for some of the bookkeeping, but it is convenient when $\text{Aut}^+(\Gamma)$ is required.

The main idea of the search for the Ford domain D is that the majority of edge cycles of D would be rather short (we hoped for only three non-EH-transformations per cycle), so that the side-pairing transformations T_ν of D would naturally arise as short words on elements T_μ with $\mu < \nu$, which again are short words on earlier transformations, and so on until the original generators $X_1 = A, \ldots, X_m$ are reached. This suggested cutting words on the original generators $\underline{X}$, which are known in advance to have an important connection with the presentation of Γ on $\underline{X}$, up into segments T in all possible ways, and considering the T as potential side-pairing transformations of $\mathcal{D}$. This was a good idea and it is implemented in the subroutine BWORDS described in Section 3. For example, the main program KNOTS3 described in Section 4 chooses the word expressing the longitude of K corresponding to the initial meridian as the one to get this treatment. In later years I added subroutines that search for certain deficiencies in a current guess at $\mathcal{D}$, and use them to build up the next guess to assist with groups Γ which are not as predictable as two-bridge knot groups. These subroutines, VXCLN and SLUIT, are so successful that one can rely on them alone in almost all cases I know about. It is not vitally important that one have a lot of advance information about the groups Γ fed to P.

Because not all discrete geometrically finite groups $\Gamma \subset SL_2(\mathbb{C})$ admit Ford domains for their action on the upper half-space model $\mathcal{U}^3$ of $\underline{H}^3$, I developed an alternative batch of subroutines that consider the action of Γ on the unit ball model $\mathcal{B}^3$ of $\underline{H}^3$. These subroutines are integrated into P so that the $\mathcal{U}$ and $\mathcal{B}$ alternatives of P rely as much as possible on common subroutines. For example, the "heavy" subroutines TEST and LELIM are intended for $\mathcal{U}$ but when used for $\mathcal{B}$ do only a little preparatory work before calling a $\mathcal{B}$ subroutine, and the heavy subroutine EDGCYC is common to both models. There was no particular reason to allow for the complications caused by an element $T \in \Gamma$ being an EH-transformation for the ball model $\mathcal{B}^3$, so the $\mathcal{B}$ version of P simply disallows EH-transformations. If such elements do occur in Γ, then for a suitable (i.e., almost any) $M \in SL_2(\mathbb{C})$ they do not occur in $M\Gamma M^{-1}$.

The final addenda to P are recent. One of these is an attempt to correct a troublesome deficiency in the system which allowed generators of Γ to be taken out of consideration entirely at an early stage, so that the subsequent calculations concerned only a proper subgroup which usually had infinite index in Γ. The subroutine RECOVR is a moderately aggressive attempt to retain the influence of ejected generators on the final result. It has helped for the particular groups Γ where this difficulty first arose, but I cannot claim that it has been extensively tested or used. The difficulty never arose for knot groups. The final addendum to P was a facility to compute $\pi\Gamma$. This batch of subroutines not only gets a

Manual for the System PNCRE

presentation of $\pi\Gamma$ but also tries to simplify it as much as possible. It seems to work pretty well for knot group images Γ, although it never seems to get presentations with much geometric meaning. In several cases it got surprising results, such as two generator one relator presentations of three-bridge knot groups, but these all checked out and I can find no fault in the answers. (Also, two generator presentations for some three-bridge knot groups were already known, but I had not paid attention to this development.)

2. PREPARATION OF THE P SYSTEM

P is being distributed as a file of Fortran subroutines (a source file) which is likely to need modification by an editor before it can be compiled by your local Fortran compiler. The compiler would produce an object module that is stored somewhere in your computer system and would be linked to the object modules of your main programs later. But to get started we have to be sure that P will compile correctly. Once the source file of P is stored in your system, examine it with an editor, and when a systematic change throughout the file is required, make the change once and for all by a global command. This will ensure that you do not overlook some cases where the change should have been made.

The distributed version of P is intended for the "fort" Fortran compiler that is available with many VAX computers. If you can use this compiler, the only changes to the source code that should be necessary concern the graph plotter (cf. below). The P file will be named pncre.f, and your object module pncre.o can be created by the command

    fort − c − V list pncre.f

Later you will access this by

    fort − o yourfilename.o − V list yourfilename.f pncre.o

At run time you will use

    yourfilename.o < yourdatafile > youroutputfile

unless the graph plotter is going to be used. Because the plotter is not available to me on my local VAX computer, I do not know how to modify this sequence to include the plotter.

Assuming that P is not suitable for your compiler, examine it using your editor. The first thing to look for is the quotation marks in one of the many FORMAT statements. If the marks appear double ("), they must be changed to single marks (') by a global edit command.

Next, compare the type of real and complex floating-point arithmetic used by your Fortran compiler with the requirement that P must use floating-point arithmetic of better than 10-decimal-digit accuracy. The seven-decimal-digit (equivalent) accuracy of single-precision floating-point arithmetic that is standard for many versions of Fortran simply will not do. The distributed version of P includes many statements like

   IMPLICIT REAL*8(A — H,O — Z)

to ensure that undeclared real variables are set to medium precision. Then real arrays are explicitly declared as REAL*8, and complex quantities are declared as COMPLEX*16. If your Fortran does not allow for the required medium-precision complex arithmetic, you will have to reprogram P by consistently using pairs of these medium-precision real numbers to replace complex quantities. (High precision refers to better than 20-decimal-digit floating-point accuracy, and although it is not used in P, it probably will be needed to get the data files for programs using P accurate to full medium-precision accuracy.) If REAL*8 and so on, is not the way to get the required precision, make the appropriate changes. The fort compiler uses such functions as

   ABS, DREAL, DIMAG, DCMPLX, DCONJG, NINT

in conjugation with this medium-precision arithmetic, and you may have to change some of these names to suit your compiler.

Beware that the subroutines PLOP, MYCRCL, and RFLCTN, which are the plotter subroutines and which have been made inoperative in the distributed version of P, do use low-precision floating-point arithmetic, as exceptions from the rest of the system. This is because their output is intended only for the plotter, and the plotter prefers or requires standard single-precision arithmetic.

The integers used in P have all been taken to be what is standard, which is INTEGER*4 for the fort compiler. This could be changed to INTEGER*2 consistently throughout if there is some reason to do so, because the integers that occur during runs of P are never very large. (This is also true for the floating-point quantities, so a compiler option which reduces the field assigned to the exponent in a floating-point number would be quite all right.)

Manual for the System PNCRE

The specific feature of P that many will find most attractive is its ability to produce handsome pictures illustrating the Ford domain $\mathcal{D}$ in U^3 that it gets. The subroutines PLOP and MYCRCL address the graph plotter by means of the subroutines

PLOTS, PLOT, SCALE, AXIS, SYMBOL, NUMBER, LINE, NEWPEN

These are supposed to be available to the load module in a file stored in the system, which was supplied by the manufacturer of the plotting machine in your installation. In the distributed version of P the calls to these eight subroutines have been converted into comments by a letter C in column 1 of each program line containing such a call. If you have no plotter, PLOP and MYCRCL do not need modification, and PLOP, MYCRCL, and RFLCTN can be deleted from the system (along with all calls to PLOP and mention of common block /CNTPP/), but do no' do this irreversibly! If you intend to use the plotter, you will have to replace the C in column 1 by a blank space for each call to the plotter subroutines listed. Also in PLOP, delete the duplicate line labeled 120 that this will cause, as well as two absurd lines between labels 10 and 20 that currently serve to avoid a warning message. Unfortunately, further changes are likely to be necessary, and I cannot say exactly what they are. Part of the problem is caused by inches versus centimeters, and part is caused by the variety of paper sizes that might be used around the world. Furthermore, different manufacturers might use different meanings for the parameters in the calls to the plotting routines; for example, the '1,-1,11' used as the last three parameters in the call of LINE just above the line labeled 130 in PLOP might not have the intended effect of putting an asterisk at each of the J points whose coordinates are stored in VX,VY without connecting them up with line segments. The plot calls in the distributed version of P are adapted to a Nicolet Zeta plotter working on 12-in.-high paper and produce a picture of $\mathcal{D}$ in a field 10-in. high. A caption is printed whose lowest line is based 0.6 in. below the bottom of the 10-in. plotting field, and the label for the highest tick mark on the vertical axis extends about 0.25 in. above it. This uses virtually all the space between the perforation lines at the bottom and top of the paper roll. The subroutines PLOP and so on, were originally developed at Southampton using centimeters and a 28-cm field for $\mathcal{D}$, and it was not hard to alter them for the equipment at Princeton, Boulder, and Binghamton. Once you have the reference manual for your plotter, the modifications are straightforward.

The two leading subroutines, BLLBL and CAYLEY, of P were once used at Bolder in the production of plots of the Ford domains in B. The actual plotting subroutines made use of software available only at Bolder and have been removed from the system. Therefore, BLLBL and CAYLEY are useless until someone writes a comprehensive plotting subroutine for B analogous to PLOP for U.

One of the practical limitations on using P is the amount of machine memory required. The system was mostly written for machines with rather limited core memory, and it has not been modified to take advantage of the much larger memory space currently available (even on the better home computers). At this writing the source file of P contains about 4600 lines and a total of about 115,000 characters. The arrays of a load module require less than 240,000 8-bit bytes. These are now considered to be rather modest numbers, so there should be no difficulty in implementing P on any computer which has good enough Fortran. Part of the reason for the modest total space required by the arrays is that we put many arrays needed only for very temporary storage into common blocks such as /SAVEIT/ and /DTIEN/, and used the space again and again for different purposes in different subroutines. This was carefully done so as not to put unnecessary restrictions on the order in which the subroutines in P can be called, but it does mean that if you need to change array sizes, you have to change the size declarations in all subroutines containing the common block in question. Actually, if the array sizes of the distributed version of P are too small for a given group Γ, you are likely to find that Γ is simply too complicated to be interesting or useful.

Beware the temptation to modify P in other ways, such as fixing mistakes! The system is cutely written, with many complicated misleading segments that look wrong but which are somehow correct. It is essential to analyze the effects of a proposed modification, with special reference to the effects in remote locations, before committing oneself to any change. If you make changes to P, adjoin a complete account of what you did to the copy of this manual on file in your computer. Then if you pass on the altered version of P to someone else, be sure that he or she gets the amended manual.

3. WRITING PROGRAMS THAT CALL P SUBROUTINES

We consider the development of a Fortran program that is to work with some explicitly defined class $\{\Gamma\}$ of subgroups $\Gamma \subset SL_2(\mathbb{C})$. This program should read from a data file the set of parameters that determine one group Γ in this class, should set up the explicit generating matrices $\underline{X}$ of Γ, and should call P to carry out decision (1), and perhaps, tasks (2), (3), (4) stated in the introduction. When the work for Γ is complete, this program would read the data for the next group, which might instead be a signal to stop. To invoke P the calling program has to set, directly or indirectly, a number of parameters that determine the action of P, then for each group it will feed the generators to P, call the search-for-$\mathcal{D}$ routines, and when the search is over, call the output routines. We consider in parallel the protocol of using the U and B alternatives.

Before P does any work, the error tolerance parameter EPS must be set. The calling program must include

    REAL*8 EPS or alt. IMPLICIT REAL*8(A — H,O — Z)
    COMMON /EEN/EPS

and will either read EPS from the data file for the run, or simply declare EPS = $\cdots$. A good choice for EPS using REAL*8 arithmetic would be 10^{-e}, with e near 8. The alternative implicit statement (if it is really needed) is a way of avoiding the declaration of simple real and integer quantities. Another common feature of the two versions of P is

    INTEGER NOG,LOA,LOG,LOR,INF,NU,NLN,NC
    COMMON /GNRTRS/NOG,LOA,LOG,LOR

This is needed for certain bookkeeping purposes.

Using the B alternative of P for a group Γ is easier than U because there are fewer options and parameters that must be set. The calling program would have

    INTEGER KSE,LO,OM,TYPE
    COMMON /VIJF/OM   /TIEN/KSE,TYPE,LO

Then before considering groups Γ by B one has

    KSE = 2

and probably at the same time also

    LO = 1

If you use B for all groups on a run, these assignments need to be made only once.

To use the U version only, common blocks /VIJF/ and /TIEN/ are not needed in the calling program. Instead, one has

  INTEGER GRPH, TYPE
    $\vdots$
  CALL SETUP(GRPH)

The GRPH in this call has allowed values zero or 1, and the call could appear as CALL SETUP(0) say. (If you do use a fixed value of GRPH here or TYPE later, these variables do not have to be declared on the INTEGER list.) The effect of SETUP is to set KSE = 1 and LO = 1 in /TIEN/, to set default values for certain parameters associated with Γ_{EH}, and if GRPH is 1, to initialize the graph plotter and set default values for parameters associated with the plots. SETUP is to be called only once per run, so if you wish to switch between U and B on one run, you should toggle the value of KSE between values 1, 2 at each switch.

U allows Γ_{EH} to be generated as shown in (1.5), where

$$A = \begin{bmatrix} 1 & A12 \\ 0 & 1 \end{bmatrix} \qquad L = \begin{bmatrix} 1 & L12 \\ 0 & 1 \end{bmatrix} \qquad R = \begin{bmatrix} i & -2iFPR \\ 0 & -i \end{bmatrix} \qquad (3.1)$$

in which

  REAL*8 A12
  COMPLEX*16 L12,FPR
  COMMON /GINF/L12,FPR,A12

and A12 is positive, L12 satisfies (1.3), and FPR, the fixed point of the action of R on $\mathbb{C}$, satisfies

$$|\text{Re}(FPR)| \leq \tfrac{1}{2} A12 \qquad |\text{Im}(FPR)| \leq \tfrac{1}{2} \text{Im}(L12) \qquad (3.2)$$

SETUP sets the default values A12 = 1.0, L12 = FPR = 0.0. The default value of A12 is so good I rarely use anything else. The default values of L12 and FPR are simply place holders, and correct values have to be set anew for each group Γ that uses L or R. The generators of Γ_{EH} are not required to belong to the generating set $\underline{X}$, although I do require that if L belongs to $\underline{X}$ then A does too and X_1 = A, X_2 = L. If any EH-generators are in $\underline{X}$ the values A12, L12, FPR should be declared now, before the impending

Manual for the System PNCRE

call of RESETG. You have to ensure that the restrictions (1.3), (3.2) are met.

There is a way to ensure that L12 and FPR satisfy the restrictions once A12 has been set. Your program has many reasons to include

COMPLEX*16 U(4),XMX(4,40)
COMMON /XMATRX/XMX

A complex matrix U in P has the meaning

$$\begin{bmatrix} U(1) & U(2) \\ U(3) & U(4) \end{bmatrix}$$

and XMX is a list of 40 such matrices that can be affected by RESETG. P assumes without checking that each of its matrices has determinant very close to (1.0,0.0) (much closer than EPS!). After A12 has been set to the correct value for Γ, you can set L12 as follows.

TYPE = 0   U(1) = 1.0
U(2) = your provisional nonreal value of L12
U(3) = 0.0   U(4) = 1.0
CALL TEST(U,INF,NU)

The value of TYPE is now 1 and L12 is safely stored. For R, which is always last, you can use

TYPE = 0   (omit this if L appears in Γ_{EH})
U(1) = (0.0,1.0)   U(3) = 0.0   U(4) = -U(1)
U(2) = your value of b of (1.4)
CALL TEST(U,INF,NU)

Now TYPE has changed to its correct value for Γ, and FPR is safely stored. This procedure can be followed either before or after the call of RESETG for generators of Γ_{EH} not in $\underline{X}$.

Next, for Γ when U is acting, one has

CALL RESETG(TYPE)

where TYPE encodes the initial description of Γ_{EH} by the scheme

TYPE	0	1	2	3	4	5
Γ_{EH}	<A>	<A,L>	<A,R>	<A,L,R>	<R>	<E>

Only these values of TYPE are permitted, and U may change them later. In the call to RESETG use the value of TYPE that corresponds to the subgroup of Γ_{EH} generated by the subset of A, L, R that occurs in $\underline{X}$. If the remaining subset has not yet been set, now is the moment. If R appears in $\underline{X}$ but A, L do not (!), then R = X_1 and you declare TYPE = 3 after RESETG (4).

The above may seem complicated and roundabout, but it is well suited to the parabolic representation of knot groups where meridians and not longitudes are naturally used as the generators $\underline{X}$. If you do not know Γ_{EH} in advance, you can start with TYPE = 5 and let U find Γ_{EH} later. However, once the value of NU (used in the calls to TEST) becomes positive, each change of TYPE is an upheaval that requires extensive reworking of the completed calculations, and the reworking might not be entirely satisfactory. There could be times when it would be best to repeat the run for Γ, making use of the extra information about Γ_{EH} from the beginning.

Beware that not declaring all the actual generators of Γ_{EH} as original generators $\underline{X}$ may produce strange effects in the description of the side-pairing generators $\underline{T}$ as words on $\underline{X}$ which the output routine BLIST will print later when the parameter LO is positive. The first line of the output for T_ν will look like

* * INDEX ν $\mu_1 e_1$ $\mu_2 e_2$ $\mu_3 e_3 \cdots \mu_n e_n$

This reads

$$T_\nu = X_{\mu_1}^{e_1} X_{\mu_2}^{e_2} \cdots X_{\mu_n}^{e_n}$$

except that EH-factors may occur in this product which do not belong to $\underline{X}$. If the program had included CALL RESETG(0) but actually TYPE = 3 (so that L and R both occur), a $\mu = -1$ appearing in the expression above refers to L and a $\mu = -2$ refers to R. Using the standard order A, L, R, the first EH-transformation not declared a generator of Γ by the call to RESETG corresponds to -1, the next to -2, and then -3.

When using U you may know at the outset an EH-transformation H $\in$ P$\Gamma L_2(\mathbb{C})$ such that $H\Gamma H^{-1} = \Gamma$ and $H^r \in \Gamma_{EH}$ for some $r \geq 2$, where H could be the glide-reflection

$$J: z \mapsto EJ*\bar{z} + BJ \qquad (3.4)$$

where $\bar{z}$ is complex conjugate, EJ = ±1.0, and BJ is complex. To use such information in the search for $\mathcal{D}$, your program should include

INTEGER ATYP

Manual for the System PNCRE

```
REAL*8 EJ
COMPLEX*16 AG12,F,BJ
COMMON /AINF/AG12,F,BJ,EJ,ATYP
```

After RESETG was called and Γ_{EH} was set up, the existence of normalizers H is declared by the scheme

ATYP	0	1	2	3	4	5	6	7
Normalizer	None	R'	A{h}	R',A{h}	J	R',J	R',A{h}	R',A{h},J

Here J is defined in (3.4) and

$$A\{h\} = \begin{bmatrix} 1 & AG12 \\ 0 & 1 \end{bmatrix} \qquad R' = \begin{bmatrix} i & -2iF \\ 0 & -i \end{bmatrix} \qquad (3.5)$$

(i.e., R' has fixed point F). The restrictions are that R' is not allowed when Γ_{EH} already contains an EH-rotation,

$|\mathrm{Re}(F)|, |\mathrm{Re}(AG12)|, |\mathrm{Re}(BJ)| \leq \frac{1}{2} A12$

$0 \leq \mathrm{Im}(AG12) \leq \frac{1}{2} \mathrm{Im}(L12) \geq |\mathrm{Im}(F)|, |\mathrm{Im}(BJ)|$

if AG12 is real, then it is positive

If you do not use this automorphism option, you can omit /AINF/ because RESETG sets ATYP to zero as the default. Of course, you could scarcely use these normalizers if you do not know enough about Γ_{EH} in advance.

For both versions of P, the calling program now declares

NU = 0

and B also has

OM = 0
NOG = 0   (RESETG took care of this for U)

The value of NU is the number of non-EH-transformations in Γ used as side-pairing transformations of $\mathcal{D}$, where T^{-1} is counted separately from T when $T \neq T^{-1}$. If, for some special reason, you do *not* want the side-pairing transformations T for $\mathcal{V}$ to have their expressions as words on the original generators $\underline{X}$ saved and printed out by BLIST later, now is the moment for

LO = 0

where LO is in /TIEN/ above. For B the setting of LO need be done only once because P never changes it, but RESETG sets the default LO = 1 anew for each Γ worked by U. There is also the allowed value LO = 2, which, in case certain troubles occur, prints out some extra information. This option is not very useful and perhaps needs further elaboration to become useful.

After making these preparations we come to the main part of the work for Γ. Your calling program should produce a sequence X_1, X_2, ..., X_m of elements of $SL_2(\mathbb{C})$ which generate Γ. We have just finished discussing the procedure for the EH-generators A, L, R, which come first, if present. To handle the rest, both versions of P need U, XMX of (3.3) and

```
    INTEGER XX(100),EX(100)
    COMMON /NWM/XX,EX
```

The most obvious way to produce the X_ν is simply to read the real and imaginary parts of the matrix entries from the data file and to convert them to the entries of U of (3.3) using DCMPLX. Once U representing X_ν is ready, use the sequence

```
    CALL TEST(U,INF,NU)
    IF (INF.LT.0) (executable statement; cf. actions to take for INF)
    NOG = NOG + 1
    XX(1) = NOG
    EX(1) = 1
    IF (INF.EQ.2) CALL WORB(1,1)
    DO 100 I = 1,4
100 XMX(I,NOG) = U(I)
```

If TEST sets INF to -1, the group Γ is not discrete and P can do nothing more for it. However, if your program has not finished reading the data file for Γ when INF is set to -1 (or -3), you will want the rest of the data for Γ to be read so that the data that follow Γ are read correctly later. The value of NOG stored in XX(1) would be the subscript ν of X_ν. If the program is using the unusual option LO = 0, the three lines XX(1) = to WORB can be omitted or bypassed. Incidentally, TEST can set INF to -3, -2, or +1, and all of these are troublesome (cf. below). You should not let one of these occur at this phase of the work, except perhaps INF = -3 using B might be hard to avoid.

Here is a list of the possible values of INF that TEST might give with their meanings.

Manual for the System PNCRE   215

INF
-1: Γ is not discrete. It has no Ford domain and P cannot get a presentation.
 0: The matrix U definitely does not yield a side-pairing transformation of $\mathcal{D}$.
 1: A new EH-transformation has been found for U which satisfies all restrictions stated above.
 2: The matrix U produced a new non-EH-transformation T which might yield a side-pairing transformation. If $T = T^{-1}$, NU has increased by one and T is stored as the NUth element of the list of non-EH-side-pairing transformations that will be used for $\mathcal{D}$. If $T \neq T^{-1}$, NU has increased by two and T is stored as the (NU - 1)th, T^{-1} the NUth element of this list. A lot of extra information for T has been stored, some for immediate use by WORB.
-2: INF would have been set to +2, but the storage arrays are full and the lists cannot be made longer.
-3: U has a shape that is not allowed by P. If U is active, $U = \begin{bmatrix} a & b \\ 0 & a^{-1} \end{bmatrix}$ where $a^4 \neq 1$. If B is active, $U = \begin{bmatrix} a & b \\ -\bar{b} & \bar{a} \end{bmatrix}$, representing an EH-rotation.

The actions to take, now or in a later phase, when the value of INF set by TEST has been recognized, are as follows:

INF
 0: No special action required.
 2: When LO is zero again no special action. But usually LO is positive and it is necessary to ensure that arrays XX, EX in /NWM/ describe U, and then call WORB.
-1: Print a suitable message and prepare to read the data for the next group.
+1: No special action when NU is zero. If NU is positive, certain arrays must be set in preparation for a call to REVISE (cf. the model procedure at the end of VXCLN). Try not to let this happen to you in one of your programs. The subroutines BWORDS, SLUIT, RECOVR, VXCLN are prepared for INF = +1. Beware that if the EH-transformation was $\begin{bmatrix} 1 & z \\ 0 & 1 \end{bmatrix}$, where z is not real and positive and if TYPE was 4 or 5, then TYPE is now 2 or 0 and A12 = $|z|$. REVISE is then supposed to "rotate" Γ (i.e., replace Γ by a suitable $M\Gamma M^{-1}$). If your generators $\underline{X}$ have not all been entered at this moment, your program will

have to rotate the rest of the generators before entering them.

-2: Your program should jump ahead to the call to LELIM, which will probably prune the lists considerably as one of its actions.

-3: Γ must be abandoned, but you can try again on a later run. If U was active, switch to B. If B was active, stay with B but replace Γ by $M\Gamma M^{-1}$ for a suitable $M \in SL_2(\mathbb{C})$.

Simply reading in the matrix entries of the X_ν is a fast way to get started with P, but it soon gets tedious. A better way to set up $\underline{X}$ is to define $\underline{X}(\Gamma)$ in terms of a set π of parameters which are to be read from the data file. For each specific group Γ the program reads π and computes the matrix entries of the X_ν by formulas. Once X_ν is stored in the matrix U, the model sequence above is to be used. A third way to get $\underline{X}$ applies when each X_ν is naturally expressed as a word on another set $\underline{Y}$ of elements of $SL_2(\mathbb{C})$:

$$X_\nu = Y_{\mu_1}^{e_1} Y_{\mu_2}^{e_2} \cdots Y_{\mu_r}^{e_r} \qquad (3.6)$$

Your program should produce $\underline{Y}$ from the data file and store the Y_μ sequentially into XMX so that the last Y goes into XMX(,40). Then for each X_ν given by (3.6), set, for n = 1 to r, XX(n) = the subscript m so that Y_{μ_n} is in XMX(,m) and EX(n) = e_n. Then

CALL SETWD(r,U)

Now the matrix for X_ν is stored in U and the sample program segment given above can be used. The array XMX of matrices is filling up from the front as this process goes on, so the Y_μ were put in at the end where they should not be overwritten. I hope the array XMX is long enough to avoid conflict.

When all the generators of Γ have been set up, the main part of the search for $\mathcal{D}$ begins. There are several strategies that one can use, one after the other, and the first of these makes use of extra information you may have in advance about Γ. Let

$$X_{\mu_1}^{e_1} X_{\mu_2}^{e_2} \cdots X_{\mu_L}^{e_L} \qquad (3.7)$$

be a word on the given generators $\underline{X}$ of Γ which is likely to be related to $\mathcal{D}$. A relator of Γ is often a suitable word. When Γ is the faithful image of a knot group, the expression of a longitude as a word on the meridians is also good, and perhaps experience with your class of groups containing

Γ will suggest other good choices. To use (3.7), read the subscripts into XX and the exponents into EX so that

$$XX(I) = \mu_I$$
$$EX(I) = e_I$$

for I = 1 to L. Then

CALL BWORDS(L,NU,MU,INF)

Here the integer MU is the number of EH-generators of Γ the program set up while NU was zero, which might be less than the number of generators of Γ_{EH}. The parameter INF is set by calls of TEST in BWORDS, and if the result was +1, the required response has already been made. If INF becomes -1 or -3, Γ must be abandoned, and if INF = -2, your program should proceed to LELIM.

There could be several words (3.7) to which BWORDS can be applied, one after the other, perhaps in a DO loop. When BWORDS is finished, the following standard procedure can be used for the rest of the work for Γ in both versions of P. The program has

INTEGER TRIES, CV, CS, TSN, TBL, CR, TN

in which TRIES is a limit on recycling through the calls to the heavy subroutines in the search for $\mathcal{D}$, and TRIES was probably read from the data file (say once for all, along with EPS and GRPH at the beginning). Also CV, CS are counting variables used with TRIES, and TSN and TBL are variables that describe the outcome of EDGCYC, the subroutine that sets up a presentation for Γ from the current $\mathcal{D}$. In particular, TBL denotes "trouble," and a value of zero means that the result seems to be good. CR, TN is the name of Γ used by the output routines PRSNTN and PLOP, and by AUTMN. Depending on the class of groups $\{\Gamma\}$, this pair would be read from the data file or computed from the data. Finally, ZR and ZI are REAL*8 variables which are somehow a useful aid for describing Γ. Your program now includes an analog to the following sequence. Certain lines are made into comments by the C in column 1 and mark addenda to the standard sequence. To make an addendum active, replace the C by a blank space.

The "(actions ...)" stand for the labels of the lines which begin the segments of your program that take the indicated actions. If your program runs B or U exclusively, the standard sequence can be simplified (e.g., for B the calls to PLOP and AUTMN should be omitted). The call to FGOS

```
      190 CV = 0
          CS = 0
          NUU = 0
      200 CV = CV + 1
C         IF (INF.NE. - 2)CALL RECOVR(NU,INF)
C         IF (INF.EQ. - 1)GO TO (actions when Γ is not discrete)
C         IF (INF.EQ. - 3) GO TO (actions for inadmissible Γ)
          CALL LELIM(NU,NLN,INF)
          IF (INF.LT.0) GO TO (actions for D being too complicated)
          IF (NU.EQ.NUU) GO TO 210
          IF (CV.GT.TRIES) GO TO 210
          NUU = NU
          CALL VXCLN(NU,NLN,INF)
          IF (ABS(INF).EQ.2) GO TO 200
          IF (INF.EQ. - 1) GO TO (actions when Γ is not discrete)
          IF (INF.EQ. - 3 GO TO (actions for inadmissible Γ)
      210 CALL EDGCYC(NU,NLN,NC,TSN,TBL)
          IF (TBL.EQ.0) GO TO 220
          CS = CS + 1
          IF (CS.GT.TRIES) GO TO 220
          CALL SLUIT(NU,NLN,INF)
          IF (ABS(INF).EQ.2) GO TO 200
          IF (INF.EQ. - 1) GO TO (actions when Γ is not discrete)
          IF (INF.EQ. - 3) GO TO (actions for inadmissible Γ)
      220 CALL BLIST(NU)
          CALL PRSNTN(CR,TN,NLN,NC,TSN,TBL)
          IF (GRPH.GT.0) CALL PLOP(NU,NLN,CR,TN,ZR,ZI)
C         IF (TBL.EQ.0) CALL FGOS(NU,NC)
C         CALL AUTMN(CR,TN,GRPH,ZR,ZI)
```

is an extra service that P can perform and causes a presentation of the fundamental group $\pi\Gamma = \pi_1(\underline{H}^3/\Gamma)$ to be printed. This presentation is written on the images in $\pi\Gamma$ of the side-pairing generators $\underline{T}$ of Γ associated with D. There is no provision in P for writing this presentation on the images of the original generators $\underline{X}$, but FGOS stores the necessary information for this by

```
      INTEGER AA(1000),EE(1000),CC(200),DD(500),R
      COMMON /TWEEDE/AA,EE,CC,DD    /FROUTF/R
```

The demonstration program KNOTS3 included with the P system contains a subroutine ORG that does the rewriting for knot group images. This also requires a second subroutine SETLG that stored information not naturally stored by P. The listings may serve as a model for rewriting the presentation in other cases.

The subroutine AUTMN uses the results for Γ together with the contents of /AINF/ to compute a Ford domain and associated presentation for the group generated by Γ, A{h},R'. It does nothing unless ATYP is positive and not 4.

There is one potentially serious problem that might arise for Γ during execution of the standard sequence. An original generator X_ν of Γ might be rejected by TEST or LELIM at an early stage because it does not correspond to a side-pairing transformation of $\mathcal{D}$, and if X_ν does not appear later in a word fed to BWORDS (and perhaps even if it does), X_ν is entirely lost from the group that P is actually working on. In this event the subsequent calculations concern only a proper subgroup of Γ that usually has infinite index. If you have reason to believe that this is happening for the groups Γ of your program, activate the call to RECOVR and the two following "IF" lines after label 200 in the standard sequence. RECOVR may cure the problem, but do not use it without good cause.

If RECOVR is not successful, you will have to revise your main program to take special action for your class of groups Γ. My own technique is to reorder the original generators $\underline{X}$ so that those with the smallest isometric spheres come first (after the EH-transformations). Let $\Gamma_0 := \Gamma_{EH}$, and suppose that for some $\gamma \geq 0$, Γ_γ has been determined. Adjoin one or two more X_ν to Γ_γ to produce the next subgroup $\Gamma_{\gamma+1}$ of Γ. Then give $\Gamma_{\gamma+1}$ the full treatment with LELIM, VXCLN, and SLUIT (but without RECOVR!) as in lines 190 to just before 220 of the standard sequence. At the conclusion, $\Gamma_{\gamma+1}$ should be properly presented, and if $\Gamma_{\gamma+1} \neq \Gamma$, the next subgroup will get the same treatment. This process is intended to produce words with large isometric spheres written on the generators with small ones. If even this technique is not good enough you will have to elaborate this idea further until it succeeds. This technique and its elaborations would also be needed in case P happens to get only the presentation of a proper subgroup of Γ whose generators do involve all the generators $\underline{X}$ of Γ.

As some solace for not including a plotting subroutine for $\mathcal{B}$ we include a subroutine which prints extra information about a Ford domain $\mathcal{D}$ in $\mathcal{B}$. The usage is

   CALL BLDATA(NU,NLN,NC)

immediately after the call to PRSNTN, in particular *before* a call to FGOS because FGOS will overwrite some arrays used by BLDATA. This will print three (long!) lists:

1. *Edge index:* indices of the spheres meeting along the edge
2. *Cycle index:* indices of the edges in the cycle
3. *Vertex index:* indices of the edges meeting at the vertex

These lists are not generally useful, so BLDATA should be invoked only rarely.

This concludes the work that P can do for Γ. A standard main program would now jump back to the place where reading the data file for the next group begins. When some quantity in the data file is a signal to stop (I often use a negative value of CR for this), the correct procedure is

    IF (GRPH.GT.0) CALL PLOP(-1,NLN,CR,TN,ZR,ZI)
    STOP

This call to PLOP closes the plot file gracefully and cannot be omitted if the plotter was ever used. The five trailing parameters of PLOP are not used in this call, so values do not have to be set for them.

We conclude with a discussion of the control parameters for PLOP that allow nonstandard options for the plots. To use any of these, the calling program includes

    INTEGER JSL,IP,JB,KC
    COMMON /CNTPP/JSL,IP,JB,KC

JSL encodes the scaling option for plotting as follows. When JSL is zero, the default value set by SETUP, the SCALE command of the manufacturer's software package for his plotter, sets the scale. When JSL is 1, the SCALE command still sets the scale, unless this would make the plot too small, in which event a larger plot is produced that still puts simple numbers by the tick marks on the imaginary axis. When JSL is larger than 1, the plot of $\mathcal{D}$ is made of maximal height, equal to the length of the imaginary axis. The second parameter, IP, of /CNTPP/ is to be left alone. The last two, JB and KC, are given the default value 1 by SETUP. You can also have JB = 0, in which case only the projections on $\mathbb{C}$ of the non-EH-edges of $\mathcal{D}$ are plotted, not the isometric circles or reflecting lines. KC arranges that PLOP plots the projection on $\mathbb{C}$ of

$$\mathcal{D} \cup A(\mathcal{D}) \cup \cdots \cup A^{KC-1}(\mathcal{D})$$

where A is the real translation of Γ_{EH}.

4. THE SAMPLE MAIN PROGRAM KNOTS3

KNOTS3 is my standard program to carry out the tasks (1), ..., (4) described in the introduction for the image of a parabolic representation of a three-bridge knot group. Let $k \subset S^3$ be a three-bridge knot with group πK. In principle, πK can be presented on three meridian generators

$$\pi K = |x_1, x_2, x_3 : r_1(\underline{x}), r_2(\underline{x})| \qquad [\underline{x} = (x_1, x_2, x_3)]$$

Manual for the System PNCRE

But the relators r_ν tend to be very long words on $\underline{x}$ and are best broken up into shorter words by using assistant generators $x_4, \ldots, x_m$:

$$\pi K = |x_1, \ldots, x_m : x_4 = r_4(\underline{x}), x_5 = r_5(\underline{x}, x_4), \ldots, r_1^*(\underline{x}^*), r_2^*(\underline{x}^*)|$$

where $\underline{x}^* = (x_1, \ldots, x_m)$. These assistant generators are not necessarily meridians, but usually I take them to be meridians corresponding to a knot projection. The distinguished meridian x_1 has an associated distinguished longitude λ_1 such that $\langle x_1, \lambda_1 \rangle$ is a peripheral subgroup of πK (and the orientations are right). This λ_1 is expressed as a word $\lambda_1(\underline{x}^*)$ on these generators, the *longitude word*. The actual data for πK that KNOTS3 uses is

$$r_4(\underline{x}), r_5(\underline{x}, x_4), \ldots, r_m(\underline{x}, x_4, \ldots, x_{m-1}), \lambda_1(\underline{x}^*) \qquad (4.1)$$

A parabolic representation $\theta: \pi K \to SL_2(\mathbb{C})$ is a representation sending a meridian x to a parabolic $x\theta$, such that the image $\pi K \theta$ is not abelian. For a three-bridge knot we can use the standard normal form

$$x_1\theta = X_1 = A = \begin{bmatrix} 1 & 1 \\ 0 & 1 \end{bmatrix} \qquad x_2\theta = X_2 = \begin{bmatrix} 1 & 0 \\ -z_1 & 1 \end{bmatrix}$$

$$x_3\theta = X_3 = \begin{bmatrix} 1 - z_2 z_3 & z_2^2 z_3 \\ -z_3 & 1 + z_2 z_3 \end{bmatrix}$$

These three complex numbers $\underline{z}$ then entirely determine $\pi K \theta = \Gamma$, and were found in an earlier project of an entirely different nature. Thus KNOTS3 and P amount to only one phase of the complete project for computing with the knot k, and the user could supply a later phase in which the results of this phase are proved correct by algorithms AD(1), AD(2) (cf. Ref. 3). A computing project with such disparate phases as this is naturally split into separate parts that are connected by using an output file for one phase as the data file for the next.

I describe the actions of KNOTS3. The first is to set up the operating environment for the subsequent work. The determining numbers

    EPS, TRIES, GRPH

are read at the beginning because they never change. KNOTS3 runs U, which is specified by SETUP(GRPH). I chose to make the graph plots (if there are any) to be as large as possible, overriding the default for this option. Three of the entries of X_2 are specified in advance because they are constant.

For each knot k seven integers are read:

CR, TN, NN, MDV, LLGT, NRTN, LAUT

(CR, TN) is the name of k (CRossing number, Table Number), and because the crossing numbers of prime three-bridge knots start at 8, a smaller CR makes a good stop sign. NN is m + 1 for the m of (4.1). For a representation θ in the standard normal form, the longitude image has the shape

$$\lambda_1 \theta = \begin{bmatrix} \varepsilon & \ell \\ 0 & \varepsilon \end{bmatrix} \qquad \varepsilon = \pm 1 \quad (\text{usually } \varepsilon = -1) \tag{4.2}$$

The number LLGT decides whether ℓ in (4.2) is part of the data for k (it is not for the sample cases included with the program). MDV and LAUT describe the automorphism of πKθ given by conjugation by A{h} as

$$h = \frac{\ell + \text{LAUT}}{\text{MDV}} \qquad \text{when MDV} \geq 2$$

If MDV < 2, there is no A{h}. The number NRTN decides whether a normalizing rotation R' is present, and if so, its fixed point F will be in the data because there is no standard way to predict F by a formula. Not all knots k have the symmetries corresponding to these automorphisms, and when they are present, the values of MDV, LAUT, and F were most likely found by inspection of a plot of $\mathcal{D}$ for k found on an earlier run.

After the call RESETG(0), which (temporarily) sets TYPE = 0, the program reads the real and imaginary parts of $\underline{z}$, and of ℓ and F if these are present. Once X_2, X_3 are set up, the words r_ν and λ_1 of (4.1) are read in, and X_4, ..., X_m, and $L \in \Gamma_{EH}$ are set up by SETWD. In Section 3 I described the use of SETWD for setting up products of matrices not in Γ, but SETWD is used in the same way for products in Γ. The matrices U for all these words are fed to TEST as described in Section 3. When ν = NN, U = $\lambda_1 \theta$ of (4.2) and TEST changes TYPE to 1 (the correct value for πKθ) while storing L12 in /GINF/ and a record of how $\lambda_1 \theta$ was converted to $L \in \Gamma_{EH}$ in another common block. Because the program tries to write a presentation for πΓ on $\underline{X}$ later, subroutine SETLG is called immediately after TEST for $\lambda_1 \theta$ to store the exact expression of L as a word on $\underline{X}$.

After this KNOTS3 is a version of the standard sequence suggested in Section 3. BWORDS is used on the longitude word in an attempt to get a good start on $\mathcal{D}$. After $\mathcal{D}$ has been found and a presentation for πKθ written on $\underline{T}$ is printed, FGOS is used to get πΓ, and ORG rewrites this presentation for $\pi\Gamma \approx \Gamma$ on $\underline{X}$ if it can. ORG will not succeed if one of the T_ν chosen by

FGOS for its presentation is not written on $\underline{X}$ because it was found using an automorphism A{h} or R'. To avoid this happening, I sometimes disable the automorphisms by putting

   LTYPE = ATYP
   ATYP = 0

immediately after the call to BWORDS. Then after calling ORG, insert

   ATYP = LTYPE

right before the call to AUTMN! Little or nothing is lost this way, because VXCLN works so well. After AUTMN the program starts a new page of the printout and jumps back to read the data for the next group.

Subroutines ORG and SETLG are included with KNOTS3. Data for the knots 8_m, m = 5, 10, 16, 17, 18, and 9_n, n = 22, 32, 42, 46, are included in the file KNOTS3DATA, which is the data file for KNOTS3. Finally, the file KNOTSOUT is the output file produced by my own version of the system, and you should compare your output file to KNOTSOUT. The differences should mainly be in the last few digits of computed floating point quantities.

REFERENCES

1. Alan F. Beardon, *The Geometry of Discrete Groups*, GTM 91, Springer-Verlag, Berlin, 1983.

2. Albert Marden, The geometry of finitely generated Kleinian groups, Ann. of Math. *99* (1974), 383-462.

3. Robert Riley, Applications of a computer implementation of Poincare's theorem on fundamental polyhedra, Math. Comp. *40* (1983), 607-632.

4. Robert Riley, Seven excellent knots, in *Low-Dimensional Topology*, Vol. I, ed. R. Brown and T. L. Thickstun, Cambridge University Press, Cambridge, 1982.

Panel Discussion

A panel met in the afternoon of Friday, April 6, to consider the issue "The Potential of Computer Algebra as a Research Tool." The panel members were Prof. R. A. Askey (Wisconsin), Prof. M. E. Fisher (Cornell), Prof. J. McCarthy (Stanford), Prof. J. Moses (MIT), and Prof. J. T. Schwartz (NYU), moderator.

Schwartz

This panel has, as usual for this sort of thing, the task of answering, in the next hour, the questions posed in the last two days. I will start the ball rolling by asking a number of questions that the panel may see fit to address.

1. What do people assume will be the impact of technology on this kind of system? This might include fast micros, large parallel computers, special VLSI chips, and other more general aspects of technology such as high-quality graphics systems.
2. What might be said of the impact of algorithmic developments, that is, development of the software side? People may wish to consider what have been the significant software developments over the past five years to get the direction and length of a vector that may point into the future. Beside this, what have been the significant systems

---

These notes were prepared by Dr. J. H. Davenport, School of Mathematics, University of Bath, England.

aspects, such as the use of graphics, or putting algebraic systems on small machines? What further algorithmic and system developments are desirable? Various things have been mentioned in the talks, ranging from better techniques for dealing with algebraic numbers to displays better integrated into algebra systems and improved numerical techniques.

3. What new application areas might develop? There might be continuing development in the area of computations for analysis and classical algebra (the MACSYMA range), computations for other areas of mathematics (e.g., topology: the combinatorial areas of algebra, such as group theory and geometry, and perhaps, symbolic logic), and nonmathematical areas such as physics, chemistry, and biology.

4. What can be expected from artificial intelligence approaches? What do people see as the ultimate possibilities and limitations for symbolic computation systems?

FIRST ROUND

McCarthy

If this conference with its present content had been held in 1965, I would have been slightly disappointed at the slow rate of progress, but now I'm encouraged by the rate of progress, so that my impatience has substantially reduced during that time. Things have, in some respects, gone quite slowly, partly for technological reasons, but mainly since it has taken a long time to develop even the people we have with the interest in computing and the mathematical capability for using it effectively. I remember that the very first efforts to use symbolic computation as a tool in group theory were very weak mathematically, and maybe not so strong in computer science either. Things have advanced quite a lot since then.

It seems to me that with regard to system questions, all the developers are beating their heads against the issue of programmability, and that somehow these systems need to be made much more programmable than they are now, so that one can try out one's ideas promptly. I myself have much interest in the following idea, which goes back to Leibnitz, who said "Let us calculate." His idea was that there would be some kind of mathematical system into which one could formulate all sorts of problems, social as well as mathematical, and calculate as opposed to argue. This proved to be a very difficult thing, and progress was extremely slow.

Leibnitz himself did not even discover boolean algebra, which is somewhat
surprising, since boolean algebra is much easier than many of the things
Leibnitz did, so one has to conclude that there was some conceptual difficulty that made that take another 150 years. Then, Boole did not discover
predicate calculus and quantifiers. Our present difficulties in applying
mathematics and logic to meet Leibnitz's goal are probably conceptual
rather than technical.

Taking things a little more narrowly, I see the application that Pro-
fessor Schwartz mentioned in his talk, that we should get to interactive
theorem proving in mathematics, and that the business where somebody writes
a program and then relies on its results with no proof that the program is
correct is something that should not be tolerated for very long. To be
specific, in the recent computer-aided proof of the four-color theorem,
the referees wrote their own program to verify that their program, written
in a different way, got the same results, and this seems to be analogous
to the referees, when somebody offers an argument for a formula by plugging some numbers in, plugging some more numbers in and agreeing that the
formula is verified for these numbers as well. If we are to rely on computers for part of the proof of a theorem, we are going to have to develop
some techniques for proving programs correct.

Askey

I would like to answer the questions by making specific comments on some
of the talks we have had. Hearn got a laugh from the audience when he
mentioned that an integral had been done by an indefinite integration routine and then checked by a table, and the table had been wrong. I was one
of the few who did not laugh, since I do not let an author use a formula
from a book of tables until the author has independently verified it,
since I take a formula in a book to be an indication that some formula
like that is true. One should not just copy formulas, and this is relevant to the designers of some present systems, who are taking things from
handbooks, because there are an awful lot of errors there, and I would
hate to see these errors propagated into systems that are shipped around
the world, and which would spread from there.

Davenport mentioned the difference between an old-fashioned algebraist and a modern algebraist. I can tell the difference by seeing how they
write a polynomial: an old-fashioned algebraist writes $a_0 x^n + \cdots + a_0$
and a modern algebraist writes it as $a_0 + a_1 x + \cdots + a_n x^n$ (modern means

that they use a computer, not that they follow the Emmy Noether style of algebra). In fact, I'm not an algebraist, I'm an analyst, and the difference is that we do not stop at $a_n x^n$, but continue and realize that power series are useful.

That raises something that was referred to by both Andrews and Chudnovsky. This is a conference on symbolic algebra, but some of us need symbolic analysis, rather than symbolic algebra, and this means more than just identities. Fateman's description of Gosper's talk as having an identity crisis hit home, for as you know, I have identities in some of my papers as well, and Littlewood said that all identities are trivial after they have been proven. But one also wants asymptotic formulas, and the systems are going to have to be expanded a lot before they can handle all that.

When it comes to indefinite integration, I'm very glad that it's automated, but I doubt that I'll ever use it. There are tables of indefinite integrals, but I've never found an integral there that I wanted and that I could not do trivially. There are many tables of definite integrals, which many scientists and some mathematicians use regularly, and we need to get to the point where we can do definite integrals as well as indefinite ones. By this I mean not only integral transforms, but multidimensional Γ and B functions that we care about a great deal, not only because we want to differentiate with respect to the parameters, and having a few numerical values is not enough to enable one to differentiate.

Gosper talked about q-hypergeometric series and referred to a special class as "vanilla." Let me describe what they are, because you all know what they are without knowing it. Many of us teach freshman calculus, including infinite series. The convergence test that students like the best is the ratio test—you take a_{n+1}/a_n, and, for all the functions that you care about in elementary mathematics, this ratio is a simple function of n. A hypergeometric series is one in which the ratio is a rational function of n. A basic, or q-hypergeometric series is one where the ratio is a rational function of q^n. Now, Euler started studying these things about 240 years ago, after he had evaluated the B function integral, and I only felt I had a relatively deep understanding of these about seven years ago, and then Gosper and a few others come up with strange things like $_2F_1(1/5)$, that I still don't understand, although my guess is that it comes out of Lagrange inversion, and is related to what polynomial equations you can solve, and so is related to algebraic geometry.

Now, let's get on to several variables, as was shown by Andrews, with his beautiful formula that represents every number as the sum of three triangular numbers, which has a single sum on one side but a double sum on the other, and then in Gregory Chudnovsky's recursion relations that he was using to produce measures of irrationality. Those are particularly simple multiple hypergeometric series, and they many not be the best type. It is clear that we are not going to be able to do these problems by hand, and we are going to have to use symbolic algebra, or analysis, to answer them. There are many different ways that we can handle these questions, and the systems being designed will be essential for some very important mathematics, and stuff that is going to be very important for applications, statistical mechanics to name but one place.

Fisher

My background is as a user of computers, and I want to use them to get an answer to a problem. The title of this is "The Potential of Computer Algebra as a Research Tool," and let me make the snobbish distinction between research and development (although I don't mind being called an engineer, and a lot of what engineers do is development). In research, the kudos go to the guy who gets there first, or gets there best. In this light, I ask myself "Is there anything here that tempts me?", and the answer is "No." Why? Then there are a variety of distinguished people who have used these systems to do various things. I agree with the point made on my right that God made the integers, and man made the rest, starting with the powers of x. It seems to me that the idea that a function is defined by its power series is one that should sink in a little more deeply, and we have already seen how useful this can be. Yet I don't find awfully good power series packages around. Rodney Baxter was mentioned earlier—he is one of the leading people in statistical mechanics to produce exact answers, and he would have been able to use some of these packages, but in fact, in view of the order to which he had to go, he wrote his own. The sellers in this market are up against the following problem: When are they going to produce a product such that it will no longer pay me and my students, who know what the problem is, to write our own system? There is a distinction here between the languages we can use, which have flexibility and are easy to understand (say, easier than an elementary analysis textbook), and the algorithms.

What is a useful algorithm? Any algorithm that gets you there is a useful algorithm, but if I want to solve a problem I am prepared to use brute force, so I don't care for a particular algorithm unless it solves a problem that cannot otherwise be solved, and we saw in Mumford's talk that things tend to go up as rather high powers, or as exponentials. I and my colleagues in London were concerned with problems of graph embedding—given a linear graph, in how many ways can one embed it in Z^n? In the first two decades of computer technology we could do better by hand. All of these problems translate into number theory in higher dimensions. In the next decade, we were testing computer hardware—divisibility checks, etc. were discovering incorrect results. I just do not see the sort of user orientation that is necessary for these things to catch on. People who have used them seem either to have known a lot about it before, or started developing their own special processes. The impression I get is that I should look out for those algorithms, like the fast Fourier transform, that will beat an exponential growth, and my response to the moderator is to ask for the algorithms that will do that. These I should know about, and they should be well packaged, and maybe they should be taught to undergraduates. Beyond that, I want an easy language, one that embodies a hierarchy of functions.

Moses

As a designer of systems, I should present the view that the existing systems have quite a few reasonable uses, many of which have not been presented here. This conference has certainly been a great success, if only by attracting such a large and diverse audience. What is also needed is another conference (and I believe that several are scheduled for this summer), where people who have used these systems reasonably successfully can come and talk to one another, and explain to the novice what sort of problems they have been able to solve and what the techniques are. That is something we have not been able to convey in this conference. At General Electric, there is a MACSYMA Users' Conference July 23-25, which is open to anyone who uses any of the other systems as well, and there is EUROSAM 84* in Cambridge July 9-11. It is always going to be the case, as it has been for the 20-odd years that I have been in the business, that there are things you want to add to these systems. Indeed, you should not expect

---

*Proceedings published as Springer Lecture Notes in Computer Science 174.

the designers to know enough about your aspects of pure or applied mathematics, or application areas, so that the users have themselves to contribute to the development of such packages.

In the past few years, there has been quite a bit of increased usage of these systems—presently several hundred sites round the country, which was certainly not true five years ago. The cost of a computer that runs a reasonable algebraic manipulation system will be between $10,000 and $20,000 in the next year or so—that's a major improvement. Referring back to McCarthy's comment, the thing that has really held us back has been the cost of the machines that run these things. When I started out, a megabyte of memory was nearly $10,000,000—in a few years it will be well under $100, and this has made, and will continue to make, a tremendous difference. Another factor, which I became aware of after talking with John Cannon, is that a large number of pure mathematicians have started using these systems, which was certainly not true a few years ago. Trying to think about why this has happened, I think that the introduction of personal computers has helped, in diminishing the barrier that people feel about using computers.

Historically, our conversations over the past couple of decades have shown that the most reasonable users have been engineers rather then theoretical physicists or mathematicians, and the reason is very simple. Engineers take a problem, and when they run up against a difficulty with symbolic computation systems, they change the problem, and so make progress. Mathematicians are different—they have a particular problem that is going to take a certain number of steps, and they can't (or at least don't) change the problem. I think this is inherent in the subject, and it is not clear that technology, in the form of expensive supercomputers, and so on, is going to give us enough of an improvement to solve problems that are truly exponential.

Although this panel is about research, most of us are in universities, and the other thing that most of us do is to teach. It has always been my goal, although the limitations of technology have prevented anyone from doing anything much about it, to try to use these systems in teaching. Indeed, even before we started the MACSYMA project in the late sixties, we had the idea of building a system that would be used in calculus teaching at MIT, and we even bought pads so that people could write, for I felt, and still feel, that the typewriter is too much of a barrier for students, even in some cases for research use, and people should be able to handwrite

their formulas. In the next few years some effort should be devoted to
this, not only at the college level, but also at the pre-college level.
This does not mean that we should revolutionize what we do—after 20 years
of thinking about calculus, I feel that there are a lot of problem solving
ideas that are introduced in classic calculus texts, even though the algorithms that are used are not ideal. If we abandon this, we may lose something that is very important to engineers and scientists. In short, I am
not looking for revolutionary changes, but for improvements in the delivery
of this material.

COMMENTS FROM THE FLOOR
Waldo Patton, Columbia
What we need is some confidence that these things work. A start in this
direction would be a suite of test programs that could be used on each
system to guarantee that they work on at least a certain set of problems.

Moses
Jeff Golden, when he used to bring out new versions of MACSYMA, would run
a battery of test programs, to ensure that the results were at least the
same as the previous version. That is a classic way of testing such systems.

Patton
Such tests should be available globally, so that one can detect not only
your own bugs but also other people's.

Jeff Grief, Inference Corp.
I work on the development of SMP, and I second both these comments. We
have a set of regression tests also, but I know that even were they 100
times bigger, they would not come close to testing all the code in the
system. Circulating sets of tests from one system to another would at
least help with the problem.

Richard Fateman, University of California at Berkeley
I'd like to respond to the previous two comments by remarking that a great
deal of testing could be done on SMP and MACSYMA if the availability of
these products, especially in the university environment, could be exped-
ited. A lot of people have been frustrated by various legal maneuvers
and ways of making money out of these programs, which have served to

separate the purpose of academic inquiry from the use of these programs. Universities are where these systems and packages originate, and the community of algebraic manipulation professionals cannot ignore that national resource that the universities constitute.

Bill Dubuque, Symbolics
This issue of verifiability and correctness is important now, and will become much more important in the future. Computer scientists have been addressing the issue of the correctness of their programs, but none of this interest has filtered through into the computer algebra world: indeed, none of the current systems have mathematically well-defined semantics for the language. Probably the most fundamental notion in mathematics is abstraction, yet none of the current systems has any idea of what it means to be abstracting something. It will be a long time before we can capture this idea, but this should be a goal.

McCarthy
It is not true that no one has been interested in the verification of such systems—I wrote a paper on proving facts about LISP programs in 1961 and have been working in that area since then, as have other people. What is true is that technology has not reached the point where verifying a system of the size of MACSYMA is a realistic proposition. The truth also is that even specifying the correctness of these interactive systems is a far-from-solved problem.

John McKay, Concordia University
Some people have mentioned the problems of communication—is the area well covered by journals?

SECOND ROUND
Moses
I wonder if Askey will recall a meeting in Madison in 1976 when we had a bet on which of two strategies would be more effective in dealing with the following problem. There is a limitation in symbolic computation that things cannot be solved in closed form all the time. The problem can be resolved in realizing that "closed form" is a relative concept—if you can't integrate a function in terms of the objects you know and love, then you should give it a new name, as has been done for a long time.

But, of course, you do not know much about this new function, and the problem arises of whether you can, via algorithmic techniques in differential algebra and algebraic geometry, semiautomatically, or even automatically, find out the properties of these functions. I consider this to be a 25-year goal, and I wrote a paper in 1972 called "A General Theory of Special Functions," and I believe this to be an important area for computer scientists and mathematicians. Askey will say that he believes in the well-known special functions, and would like more progress on them.

Fisher

Price is important. The slide rule was around for a long time and did a lot for engineering since people could afford one each. I don't reckon that $10,000 makes real sense, even though some departments are getting richer (and others poorer). Once we are below $1000, which is where the H-P calculators first came in, there will be real progress. I would like to know more about the possibility of a stripped-down system that would do the elementary things, which would still be of interest to me. Pricing is a complex issue, but we should concentrate on the lower end.

I'd also like to mention displays, which relates to teaching. From what I have seen, I would be against letting anyone learning the subject use an algebra system, since you ought to multiply out your own matrices, even if they are all 2-by-2 or 3-by-3. The display aspect is crucial for learning and understanding. If I were using these systems, I think I would want to see functions plotted this way or that way practically all the time. When I look at mathematics graduates, I find that they cannot think geometrically—they do not know whether a function is going up or down, what the difference is between a cusp and a maximum, and in general, that they can't think visually. So the question is whether you can get good displays down to a price that people can afford. A large proportion of working engineers and mathematicians do a large fraction of their thinking visually, and computing could make this go further.

I am very encouraged by the renaissance of this area. I would urge the systems people to have definite problems in mind, even if they can't have them all in mind, because they can then say "My system does problem X." This may lead to more specialized systems, but is that a bad thing? CAYLEY, for example, does a lot for group theory, and that is good, even though I personally found group theory a pain in the neck.

Askey

After Wolfram's talk, Kahane said that we cannot ignore elementary calculations just because there are pocket calculators, since they do not do everything right. You should hear his lecture on pocket calculators. I will mention two problems in the area that Wolfram hopes to be able to automate. Walter Gautschi was computing confluent hypergeometric functions by a standard recursion relation coming from the continued fraction, and after five or six terms it had settled down, and the next four were very small changes. Unfortunately, it had settled at the wrong place, and at 20 terms it changed and settled down somewhere else. You do need to worry about errors, and you do need to get error estimates that will tell you whether you have finished or not. The same thing is true of gaussian quadrature. There was a paper in a chemistry journal that was doing integration from -x, to x, with respect to $e^{-|x^3|}$. The chemists gave the location of the zeros of the orthogonal polynomials and the Coates numbers to 15 places of decimals, and checked it by doing a few functions that they knew how to integrate. Unfortunately, as Gautschi* points out, this test is useless, and their table has at most two accurate places anywhere. You have to think with computers, just as you have to think with tables.

If I look at Knuth's book, Section 1.2.6, he says that it is essential to learn to handle sums of products of binomial coefficients, and he wrote down six rules for handling them. What he did not know is that there is an algorithmic way of handling these, and that his first five equalities are exactly the same. The method is called hypergeometric functions. This illustrates the fact that combinatorics used to be called combinatorial analysis, and the people currently doing it ignore analysis at their peril.

According to my recollection, Moses said that differential algebra applied to differential equations would be the way to study special functions. Since the most important special functions are related to difference equations rather than to differential equations, I win the bet. But that may no longer be the case. Here is a research problem, where I think a computer plus a lot of hard thinking and some algebra (orthogonal groups, etc.) might play a crucial role. Chandrasekhar published last year a book on black holes. He said it is the most complicated thing that he has ever done. He deposited 6000 pages of calculations relating to the solutions

---

*BIT *23* (1983), 209-216.

of the differential equations in the library of the University of Chicago. Only the results went into the book. The solutions of these differential equations have properties they have no right to, and the question is why. They are not hypergeometric functions, yet the identities that he has are not just the trivial ones that come from Wronskians. It has taken a great mathematical physicist 10 years to get somewhere with these equations, and it would take anyone else two or three years to understand what he has done, but it is worth it.

Graphics was mentioned. Janke and Emde has great graphs, which have not been touched since,† and it is a scandal. We should have graphics packages that should generate the sort of pictures we can put in books. There's a lot of interesting mathematics* that can be done this way—don't get me wrong. I just want to see them pushed further.

McCarthy

I have been asked to say something about artificial intelligence. In the earliest work in computer algebra, in particular, integration, we tried to model the methods that were taught in calculus courses, and we achieved some moderate success. Much greater success has been achieved by more systematic algorithms. It seems to me that this is a temporary phenomenon. When artificial intelligence makes more progress, we will be able to get real help out of it. When will I be able to use a computer to amplify my intelligence, to the point where I can program it to explain things to me? I really mean explain them, not merely verify some conjecture or perform some computation. It seems that we are not very close to that.

If we look at the current state of artificial intelligence, the large technology is the rule-based system. These systems are to some extent intelligence degraders, in the sense that although they may be useful, the work that went into inventing the rules is of a higher intellectual level than the work being carried out by the system. To take one example, many of the medical diagnosis programs do not predict what the treatment will do, but the people who wrote the rules did so in order to say what the treatment should be. It will be slow work to overcome this situation.

---

*See (5) and (9) in the speaker's paper in these proceedings.

†*Note added in proof:* But see the thesis of F. Richard, University of Strasbourg, Sept. 1988.

Artificial intelligence will be with us until it is solved. Those who say that it has not reached human level in 30 years, so we should give up, are analogous to those who might have said in 1900 that we do not know the structure of inheritance, so should give up on genetics.

Index

ALGEB, 148
Algebraic
  functions, 133
  numbers, 133
Artificial intelligence, 236
Automatic formula manipulation, 184

Bailey's lemma, 103
Baker's method, 17, 37
Beta integrals, 121

Canonical representations, 130
CAYLEY, 89
Celestial mechanics, 111
Characters, 94
Chebotarev theorem, 147
Complete intersection, 88
Computational complexity, 183
Computer algebra, iii, 124
Constant term identity, 124
Constructive algebra, 135
Convolution, 183
Cramer's rule, 135

Dense representation, 131
Differential algebra, 1, 5, 75
Diophantine approximations, 1, 2, 34
Discrete group, 199
Divisor chain condition, 136
Dyson's conjecture, 122
Dyson-Andrews conjecture, 123

Edge cycle representation, 199

Factorization
  modular methods, 139-140

[Factorization]
  p-adic methods, 139
Ford fundamental domain, 199
Fourier transforms, 183

G-functions, 55
Galois group, 145
Gaussian
  elimination, 134-135
  polynomials, 98
Gauss sums, 126
gcd domain, 136
Göllnitz's theorem, 105

Hard hexagon model, 97-98
Hilbert
  modular equations, 83
  modular group, 85
Hyperbolic space, 199
  forms of finite volume, 200

Integer, 129
Invariants, 146

Jacobi sums, 126
Jacobi's triple product identity, 124

Kolchin problem, 4

Mackdonald-Morris identity, 124
MACSYMA, 83, 124
Matrices, 133-135
Mock theta functions, 102
Modular, 130
Moon, 111, 188

Norman's technique, 142-143

p-groups, 94
Pade approximations, 2, 4, 47
Partitions, 97, 105-106
Polynomial remainder sequence, 138
Polynomials, 130-132
Power series, 141

Radix, 130
Ramanujan, 40
Rational
　functions, 132
　numbers, 130
Rectangular transforms, 188
REDUCE, 83
Ring class field, 87
Roger-Ramanujan identities, 97-98, 101-104

SCRATCHPAD, 1, 14, 98-105, 107-108, 184

Selberg beta integral, 121
Semidirect product, 90
Set transitive, 147
Singular moduli, 85
$SL_2(\mathbb{C})$, 199
SMP, 1
Sparsity, 131
Statistical mechanics, 97-98
Steiner system, 146, 149

Tchirnhaus transformation, 147
Thue equation, 11

Unique factorization domain, 136

Verification, 232

Weber
　modular function, 84
　theorem, 87